Biochemical & Medicinal Chemistry Series

Series Editor

JOHN MANN

Professor of Organic Chemistry

Titles in this series

Biochemical & Medicinal Chemistry Series

Antiviral Chemotherapy

RICHARD CHALLAND

Head of Anti-Infective Chemistry Section at the former Wellcome Research Laboratories, Beckenham, UK

and

ROBERT J. YOUNG

Senior Chemist, Anti-Infective Chemistry Section at the former Wellcome Research Laboratories, Beckenham, UK

Spektrum

Copublished in the United States with
UNIVERSITY SCIENCE BOOKS

Spektrum Academic Publishers
33 Beaumont Street, Oxford OX1 2PF

Distributed by
W. H. Freeman at Macmillan Press Limited
Houndmills, Basingstoke, RG21 6XS

Copublished in the United States with
University Science Books,
55D Gate Five Road, Sausalito, California 94965
Fax 415–332–5393

100742699T

British Library Cataloguing in Publication Data

A catalogue record for this book is available from the British Library.
ISBN 1–901217–03–5

Library of Congress Cataloging-in-Publication Data

Challand, Richard, 1946– :
Antiviral chemotherapy / Richard Challand and Robert J. Young
p. cm. —(Biochemical and medicinal chemistry series)
Includes bibliographical references and index
ISBN 0–935702–94–6 (softcover)—ISBN (invalid) 1–901217–03–5 (softcover)

1. Antiviral agents   I. Young, Robert J. (Robert John). 1963–
II. Title.   III. Series: Biochemical and medicinal chemistry series.
[DNLM: 1. Virus Diseases—drug therapy.   2. Antiviral Agents—
therapeutic use.   WC 500 C437a 1996]
RM411.C43 1996   616.9'.25061—dc20   DNLM/DLC
For Library of Congress   96—44 CIP

Design and DTP by Pete Russell, Faringdon, Oxon
Printed by Bell and Bain Limited

Contents

Preface

The science of virology as a comprehensive discipline has only emerged during the last 40 years or so. Its development has followed closely upon the exciting discoveries made in sub-cellular biology and biochemistry such as the nature of the genetic code (Watson, Crick and Wilkins), the function of ribosymes (Cech and Altman), the characterization of the first viruses (Enders, Weller and Robbins) and the determination of their structure by crystallographic means (Klug), all of which have attracted Nobel prizes. The increasing understanding of the role that viruses play in human disease inevitably led to the exploration of potential treatments, often, in the early days, on an empirical basis. Vaccination against viral conditions had become the norm in many cases, even though there had been no knowledge of their causes; indeed, it was predicted that this would be the only means by which they could be treated by virtue of the virus's biochemistry being so closely linked to that of its host cell. Fortunately that prediction proved to be simplistic and a number of effective antiviral agents are now generally available, notably through agents which affect nucleic acid biochemistry, an area in which the work of Hitchings and Elion also led to a Nobel prize. The advent of HIV infection and the unprecedented scale of multidisciplinary investigations into its treatment, has hastened advances in virology in general.

In 1994 the market for antiviral chemotherapeutics was worth about £1.8 bn, of which about half was attributable to a single drug, acyclovir. This represents a phenomenal expansion in a period of less than 15 years and underlines antiviral chemotherapy as one of the major success stories of modern medicinal research. Current predictions are for this market to quadruple within the next ten years, in response to an ever-increasing number of effective agents with diverse modes of action and viral targets.

In this book the history of the discovery of antivirals is covered briefly, leading on to the development of more and more sophisticated and selective treatments; reflections on the increasing understanding of viral biochemistry and its difference from that of host cells. The intention is not to be comprehensive but to provide a general introduction to antiviral chemotherapy from the point of view of a reader already adequately versed in chemistry or biochemistry. Although the emphasis is upon the biochemical mechanisms of action of small molecules and their development as drugs, the applications of vaccines, interferons and oligonucleotides are also covered briefly.

A reading list directing the reader to more comprehensive coverage of the topics referred to in the main text is at the end of the book, along with a reference list of contemporary marketed or experimental antiviral drugs and a glossary of terms in quick reference format.

We would like to acknowledge the assistance of Prof. John Mann, Dr Peter Collins, Dr Brendan Larder, Dr David Stammers and Dr Paul Smith in the preparation of this manuscript and Dr. Allen Miller and the former Wellcome Foundation Ltd for logistical support throughout. We also thank Michael Rodgers, Alice Nelson and the staff at Spektrum, Jane Templeman for careful illustration, Pete Russell for his triumphs in the face of technological adversity and finally Julia and Claire for their patience and encouragement.

1 General introduction

The discovery of viruses

One of the major reasons why the average life expectancy used to be shorter than that to which we have become accustomed in the developed world, was the prevalence of infectious diseases. The mechanisms of many of these infections began to be properly understood during the nineteenth century. Almost concurrently, it was demonstrated that some protection against such diseases could be gained by immunization techniques, even though the mechanism of action was not properly understood. The characterization of bacteria was rapidly expanding by the 1920s and during the Second World War, considerable therapeutic benefit was derived from the treatment of such infections by penicillin. This led to the widespread belief that modern medicine would soon provide agents capable of treating all infectious diseases.

Yet there remained a large number of conditions which did not fit the pattern. Not only were the newly discovered 'antibiotics' ineffectual in alleviating symptoms, but it was also impossible to detect the pathogenic agents by microscopic means or by culturing samples taken from patients. The first clues as to the nature of the pathogenic agents which were not susceptible to antibiotics came from the study of plant diseases. In the 1930s the first virus was isolated from diseased tobacco plants and given the name tobacco mosaic virus (TMV) after the effect it had on the leaves of the plant. It was subsequently shown that the particles of TMV, which were in the form of rods of variable length containing both protein and nucleic acid material, were not viable on their own account, but were capable of reproducing themselves when inserted into suitable cells. Soon afterwards viruses were discovered which infected bacteria—the bacteriophages—studies of which showed that the nucleic acid is essential for replication but most of the protein is not. Subsequently, viruses which infect mammalian cells were discovered and demonstrated to be the aetiological agents behind many of the infections which were not susceptible to antibiotics.

The beginning of modern virology

During the 1940s and 1950s, advances in cell biology and the development of sterile cell culture techniques, themselves owing much to the application of the new antibacterials, enabled human viruses to be studied in detail in the laboratory. The work of Enders, Weller and Robbins with poliovirus, grown under controlled conditions in cell culture, enabled virus particles to be harvested, purified and crystallized, thereby enabling the morphological structure to be determined by using electron microscopy. The poliovirus particle (Figure 1.1) was shown to be very small, not much bigger than the largest known protein molecule (about 30 nm), but with a regular icosahedral structure. The nucleic acid in the core was exclusively RNA and the protein coat was shown to exhibit a strong immunological response.

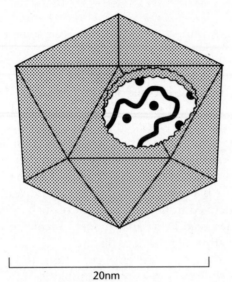

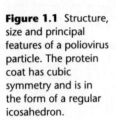

Figure 1.1 Structure, size and principal features of a poliovirus particle. The protein coat has cubic symmetry and is in the form of a regular icosahedron.

20nm

The nature of viruses and their link with human disease was thus firmly established; subsequently, many more viruses have been, and continue to be, discovered. Their structure, occurrence and effects show great diversity: some cause profound changes in the cells they infect and lead to serious disease; others seem to have very little overt effect on their host and some may even be beneficial. The remarkable thing is that most viruses are very selective in their infectivity; indeed, not only are many highly specific for a particular host, but also for particular organs or cell types within that host.

Vaccination against viral diseases

The development of vaccines to combat viral diseases was not a consequence of the discovery of viruses. Protection against smallpox had first been demonstrated scientifically by Jenner in 1796 by inoculation with cowpox, although the true origin of such techniques is certainly much older. However, a better understanding of the mechanisms involved has led to the development of many successful vaccines against viral conditions; indeed smallpox has been totally eradicated due to vaccination and polio is effectively controlled by an inexpensive programme. Such preventative methods are the most effective and economic way of combating most viral conditions. They are, however, only preventative measures and cannot be used to treat diseased individuals. Whilst this book does not cover immunology in depth, reference is made to the applicability of vaccination and to the underlying mechanisms in appropriate cases.

The advent of antiviral chemotherapy

In the early days, the effectiveness of many vaccination programmes and a long-held dogma about the implausibility of discovering effective antiviral drugs prevented much research into such agents. However, some efficacy was reported against smallpox by the prophylactic use of methisazone (Figure 1.2), followed by several poorly selective nucleoside molecules with some topical utility against Herpesvirus conditions. Subsequently, acyclovir (Figure 1.2) emerged as the first truly effective and selective antiviral agent. The discovery of acyclovir owed much to an understanding of the fundamental biochemistry of viral replication, in particular to nucleic acid synthesis, a field in its infancy in the 1950s. These concepts—especially the genetic code and the mechanism of protein synthesis—meant that the biochemistry of viruses could be understood and, of more importance, contrasted with the host cell mechanisms. Subsequently, the techniques of modern molecular biology and the ability to perform *genetic engineering* has had a profound effect. The gene sequences of viral nucleic acids can be determined; individual proteins from them can be expressed, thus their function and the nature of their substrates determined. Likewise, sophisticated assay systems have been developed, which have been invaluable in the search for potential new antiviral drugs. Unlike some forms of drug therapy, which often have roots in alchemy and folklore, modern antiviral drugs have been designed and refined through an understanding of the fundamental processes involved. The aim of this book is to provide insight into the way that such drugs have been discovered by rational approaches.

Figure 1.2 Structures of methisazone and acyclovir.

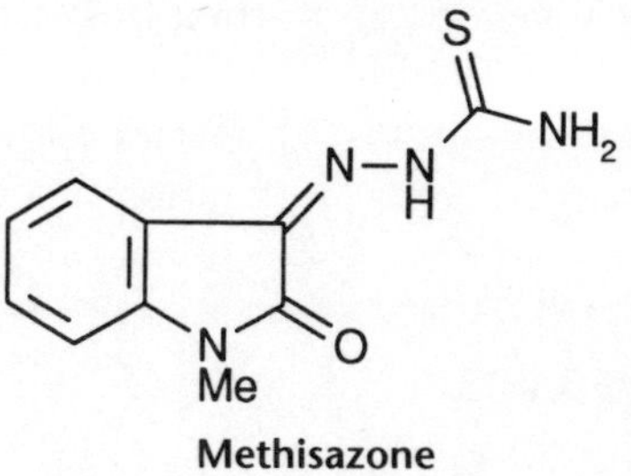

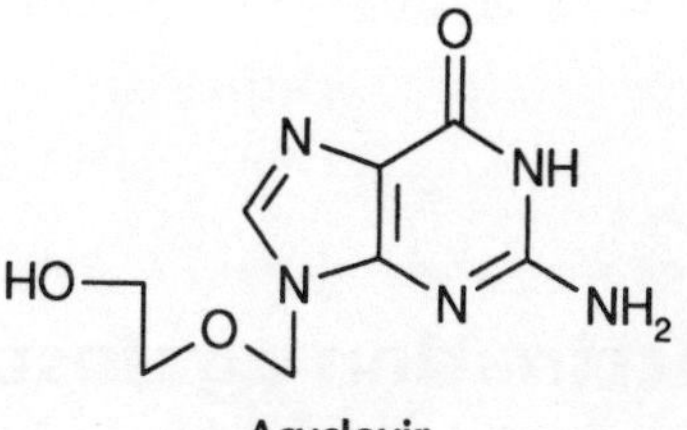

Methisazone

Acyclovir

As this introduction has suggested, antiviral chemotherapy is still a young science, especially when compared to vaccination techniques or other branches of chemotherapy. Still, an ever-increasing number of viruses have been studied in assay systems and candidate drug molecules identified against many of them. It is true to say that, due to economic pressures, the viral targets which have received most attention for potential chemotherapy are, sadly, those mainly afflicting the developed world. By contrast, the availability of inexpensive vaccines against many viral infections has provided widespread benefit for all mankind.

An important distinction that must be emphasized here is the clear difference between viruses and other microbial pathogens. The frequent confusion displayed in the news media and even by some professionals, between a viral disease and a bacterial or fungal infection has no excuse. Viral infections cannot be treated with antibiotics nor can antivirals be used to treat other microbial infections: neither the terms virus and bacterium nor their treatments are interchangeable. Having made this distinction, it should be appreciated that opportunistic infections of bacteria and fungi can occur as a secondary event following a viral disease and in such cases might require their own, independent, treatment.

Organization of this book

The first three chapters of this book give an introduction to viruses through their occurrence, biology and biochemistry. An awareness of the concepts introduced in these chapters is fundamental to an understanding of the rationales used in the design of chemotherapeutic agents. A chapter on nucleic acids, particularly nucleosides, follows, which deals with such molecules, the most important viral chemotherapeutic targets, from the point of view of medicinal chemists. In the main body of chapters, chemotherapeutic approaches to important individual groups and families of viruses are discussed, with a closing chapter which highlights potential areas for future progress in the field.

2 Introduction to viruses

What are viruses?

A virus is a subcellular organism which, as a discrete entity, is inert or biologic-ally dormant, but once it has entered a living cell is capable of subverting the biochemical apparatus of the cell to ensure its own replication. The range of entities which is encompassed by this description is very large; some viruses are closely related to others in structure whilst others have highly diverse morph-ology (Figure 2.1)

The earliest system of classification was based purely on the viral morphology and the nature of its genome, which is still useful and will be described. Latterly, viruses have also been classified within species according to the genetic make-up of their proteins, resulting in a sub-system of strains and serotypes. First it is nec-essary to learn the basic features of virus structure.

Structure of viruses

In the simplest terms a virus consists of a genome composed of nucleic acid, either in the form of RNA or DNA, which is packaged along with proteins inside a protective coat, which may also be composed of proteins but includes, in many cases, lipids as well. For a small RNA virus such as human rhinovirus, responsible for many colds, the basic features are quite simple. The virus particle consists of a strand of RNA (the **RNA genome**) and a very limited number of viral enzymes (**non-structural proteins**) packaged inside a protective protein coat. The internal material of the virus is termed the core. The coat is often, but not always, in the form of a regular icosahedron and is made up of many ident-ical subunits (**capsomeres**) each made up of a small number of discrete struc-tural proteins. Rhinoviruses are thus able to construct themselves with a minimum of genetic material, as very few individual proteins need to be encod-ed in the genome. The **structural proteins** possess the capacity to self-assemble without the assistance of enzymes or other outside influence into a ball identical in size and shape to a mature virus particle, such a structure being termed a **capsid**. This process occurs in infected cells very readily and when followed by insertion of the core results in **reassembly** into mature virus particles (**virions**). Viruses which are constructed in this way possess **cubic symmetry** which is inherent in the economical way in which they use their raw materials. The

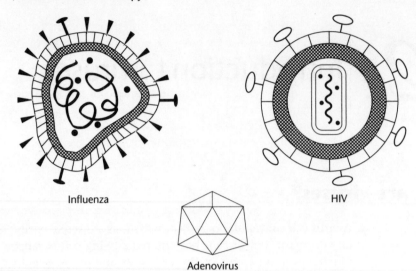

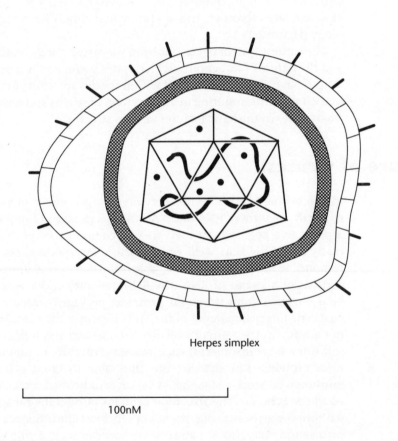

Figure 2.1 Typical sizes and shapes of virus particles. Note the preponderance of regular shapes which enables the macro-structure of the virus particle to be built from simple repeating units.

Influenza

HIV

Adenovirus

Herpes simplex

100nM

number of capsomeres required to build up a complete icosahedron varies according to the type of virus; generally speaking the larger the number, the larger and more complex the virus. For example, rhinoviruses have 60 identical capsomeres per virion, whereas adenoviruses have 252 capsomeres which are not all identical, some containing different proteins from others. Adenoviruses are distinguished also by having a genome consisting of DNA, rather than RNA. It possesses a **DNA genome**.

Whether made of RNA or DNA, the genome may exist as a single molecule of nucleic acid in the virion (**single stranded** or **monocistronic**) or several discrete pieces may be present (**multi-stranded** or **polycistronic**). The gene may also be in the form of a **positive strand** or a **negative strand**, the precise meaning of which will be explained fully in Chapter 4. Examples of viral genomes are shown in Figure 2.2.

(a)

Figure 2.2 Genomic maps of typical viruses. (a) Poliovirus RNA genome encodes only about eight proteins which are expressed as a single chain. The viral protease subsequently cleaves it into into the required fragments. (b) Hepatitis B possesses a circular double-stranded DNA genome. The proteins are derived from overlapping reading frames. An economical use of the genome!

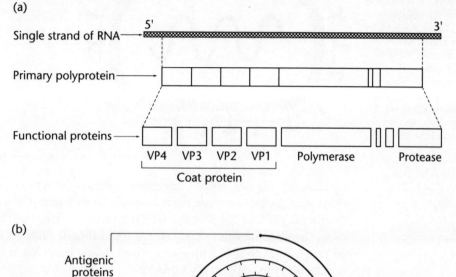

(b)

Other viruses, such as tobacco mosaic virus (TMV) are capable of assembling their capsids with a single structural protein. TMV particles are in the shape of rods consisting of an RNA genome packed into a helix in association with the single structural protein. Every molecule of the structural protein is therefore in an identical environment and the particle has **helical** symmetry, an even more economical use of raw material. Helical symmetry is more common amongst enveloped viruses.

In the measles virus (Figure 2.3) the core conforms to the TMV pattern in that it possesses an RNA genome complexed with multiple copies of a single nucleoprotein. This core is surrounded by other protein called the **matrix protein** and the whole is surrounded by an **envelope** consisting of a lipid bilayer containing **glycoproteins** of viral origin. These glycoproteins are sometimes referred to as spikes and play a significant role in the way the virus recognizes its target cells. The proteins and glycoproteins are encoded by the viral genome but the phospholipids of the envelope itself are derived from the host cell membrane.

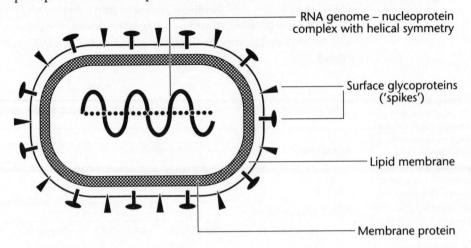

Figure 2.3 Measles virus is a typical RNA enveloped virus. The genome is complexed with nucleoprotein into a nucleocapsid in the core. The membrane Protein and surface glycoproteins are associated with cell-derived lipids which form an envelope around the core.

RNA genome – nucleoprotein complex with helical symmetry

Surface glycoproteins ('spikes')

Lipid membrane

Membrane protein

Likewise, viruses with underlying cubic symmetry may also have lipid envelopes. Herpes simplex is an example from a group of viruses, discussed in Chapter 6, with a DNA genome which is relatively large and biochemically complex. The genome can encode about 100 different proteins and the virus has a protein coat as well as a lipid envelope. Yellow Fever Virus is an example of an enveloped virus with an RNA genome.

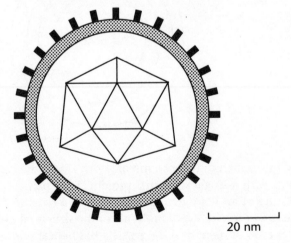

Figure 2.4 Yellow fever virus showing the regular icosahedral symmmetry of the capsid, surrounded by a lipid envelope bearing virally derived surface proteins

20 nm

The relatively recently discovered human immunodeficiency virus (HIV) is an example of a **retrovirus** which does not closely resemble any of the types discussed above. HIV possesses an approximately spherical lipid envelope with spikes enclosing a core containing the RNA genome and non-structural proteins. The major point of difference from other common viruses lies in the biochemistry of its replication cycle, the chief aspect of which is that the viral genetic information becomes incorporated into the DNA of the host cells and the latter become **transformed**. The main families of RNA and DNA viruses involved in human disease, with examples of each type, are given in Table 2.1 and Appendix 1.

Table 2.1 The structures of some DNA and RNA viruses

DNA Viruses

Genome:	Single Strand		Double Strand	
Capsid Symmetry:	Cubic	Cubic	Cubic	Irregular
Enveloped:	?	Yes	No	Yes
Families:	Parvoviruses	Herpesviruses (Herpes simplex) (Cytomegalovirus) (Varicella zoster)	Adenoviruses	Poxviruses (Smallpox)
			Papovaviruses (Papilloma)	Hepadnaviruses (Hepatitis B)

RNA Viruses

Genome:	+ve Strand		Helical	
Capsid Symmetry:	Cubic		Helical	
Enveloped:	Yes	No	Yes	No
Families:	Togaviruses (Rubella)	Picornaviruses (Rhinovirus) (Polio)	Coronaviruses	Rigidoviruses (Tobacco mosaic)
	Flaviviruses (Yellow Fever) Retroviruses (HIV)			

Genome:	–ve Strand		Double Strand	
Capsid Symmetry:	Helical		Cubic	
Enveloped:	Yes	No	Yes	No
Families:	Orthomyxoviruses (Influenza) Paramyxoviruses (Measles) Rhabdoviruses (Rabies) Arenaviruses (Lassa Fever) Filoviruses (Ebola)			Reoviruses (Rotavirus)

The antigenic behaviour of viruses

The basis of the mammalian immune system is that foreign chemical entities called **antigens**, usually fairly large molecules, stimulate the production of

antibodies, which are proteins produced by blood lymphocytes which bind specifically to a particular antigen by forming a complex with it (Figure 2.5). In the case of viruses the antibody binds to specific protein sequences on the exterior of the virion and neutralizes its biological effect by masking it from its environment and preventing penetration into cells.

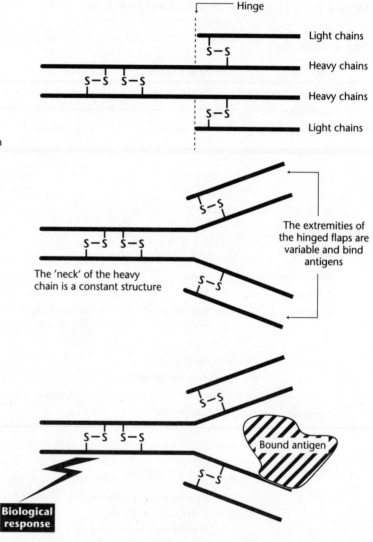

Figure 2.5 Diagrammatic representation of antibody response. The immunoglobulin molecule light chains adapt themselves to bind the antigen, which once attached triggers the constant region into eliciting the biological responses which constitute the immune reaction.

All viruses which infect animals contain molecular surface features which are capable of eliciting an immune response in the host and this property can be used as a system of classification. For non-enveloped viruses this is inherent in the protein sequence from which the exposed part of the capsid is made, whilst for enveloped viruses regions of the protein spikes embedded in the envelope possess the same property. In some cases these regions are well characterised, an example being those of the surface proteins of influenza virus, haemagglutinin (HA) and neuraminidase (NA), both of whose structures are accurately known from X-ray crystallographic studies. The ability of the host immune system to raise antibodies to viruses constitutes the first line of defence in resisting infections. The process results in many infections, such as colds, being minor and self-limiting and others, such as cytomegalovirus infections, being essentially free of symptoms in immunocompetent individuals.

Antibodies to viruses are usually highly specific for species with a particular molecular surface feature and will only bind to and thus neutralize individual virions displaying that same feature. This property can be made use of under laboratory conditions for identifying and classifying viruses. Some viruses are very easy to neutralize in that antibodies raised from one stock of virus will readily bind to other stocks and prevent them from causing an infection. However in other cases, because of random copying errors in the replication of their genes, a process known as **mutation**, the viral surface proteins undergo systematic variations in their peptide sequences and this leads to a situation in which antibodies raised to the original virus become ineffective against the mutated virus. The role played by gene mutation in the development of resistance will be covered fully later.

Accordingly most viruses which have been studied from an antigenic point of view have been classified into **serotypes**. Serotypes are versions of that particular virus which are identical in all respects except for those portions of their surface proteins which elicit antibodies. More than a hundred different serotypes of rhinovirus have been characterized. Conversely, there are only a few invariant serotypes of poliovirus. Viruses of the same species which have modified their proteins in any manner are usually termed **strains**. Strains of a virus which have differences in the core proteins but not in the surface features may not therefore be antigenically distinct and would still belong to the same serotype. Influenza virus occurs in two distinct forms called 'A' and 'B' strains several of whose proteins have detail differences and result in a different response to chemotherapeutic agents.

The response of the host immune system is clearly one of the most important lines of defence in combating viral infections. Symptoms may still arise however, in situations where the pathogenic effects of the virus may reach a serious level before the immune system has had time to deal with it, or where the virus mainly replicates in a part of the body which is not accessible to the immune system.

Problems also arise in hosts whose immune system is compromised in some way so that the natural ability to fight infections is reduced. Such hosts may be patients who are receiving immunosuppressant drugs as part of cancer treatment or to help with accepting a transplanted organ; or they may be those who have a diseased immune system resulting from leukemia or by infection with HIV. These patients, **immunocompromised (IC) hosts**, are assuming vastly increased importance in modern medicine, not least from the larger number of serious infections, including viral diseases, to which they become susceptible.

Vaccines against viral diseases

In theory, once antibodies have been raised to a particular pathogenic virus, infected individuals should not be susceptible to a similar infection again; they become **immune** to the infection. Even before the exact nature of viruses was known it was recognized that an immune state could be achieved in an individual by artificial means, most often by injecting 'dead' viral material or 'attenuated' (non-pathogenic) live virus into the circulation, a procedure termed **vaccination**. The 'dead' material contains the antigenic protein sequences of the viral coats but not the viable core material which would enable the virus to replicate. Vaccination programmes have been very effective in controlling polio, measles, mumps, rubella, yellow fever, hepatitis B and rabies and have completely eradicated smallpox. Generally speaking vaccination is a cheap, efficient and effective means of preventing disease and has brought about untold improvements in the quality of life in the modern world. It is important to appreciate that vaccination is a preventative measure and is of virtually no use once an individual has become infected because the virus will by then have already penetrated the cells.

Not all viral diseases can be controlled in this way. The technique has clear advantages for those viruses which only have a limited number of invariant pathological strains but for viruses with many distinct antigenic forms and for those which are capable of rapid alteration of their antigenic state problems arise which can be best illustrated by reference to the common illnesses caused by rhinovirus and influenza.

Rhinovirus exists in about 100 immunologically distinct serotypes which circulate intermittently in the population. In theory it would be possible to develop a vaccine for each of these. But in the first place it is not possible to predict which particular serotype may be predominantly circulating at any one time and secondly the time scale on which the vaccine could be produced is such that another serotype could have then assumed predominance. Natural immunity to rhinovirus develops very quickly in any case so that the predominant serotypes present in the population at any one time are always shifting. Indeed this effect is amplified in the modern world because of the prevalence of mass travel and consequent rapid dissemination of new serotypes. It is just not practical to carry out a vaccination programme in such circumstances.

The antigenic determinants of influenza are located on the molecules of haemagglutinin (HA) and neuraminidase (NA) which make up the surface spikes of the virus. Infection, though having more serious symptoms than rhinovirus, is again self-limiting and immunity to that particular infectious strain is quickly established in the host. During the viral replication cycle the antigenic regions of HA or NA may gradually change so that after a period of time the virus becomes unaffected by existing antibodies in the population. Thus a new strain of the virus emerges which can infect unprotected hosts and an epidemic starts. Occasionally HA or NA undergo a major mutation by a process of gene recombination and virus emerges with a completely changed antigenic profile. In recent years this happened in 1947 (H_1N_1 strain), 1957 (H_2N_2 strain) and 1968 (H_3N_2 strain) and the result is always a major world pandemic of influenza with many deaths and very high levels of morbidity. A change also occurred in the late 1970s but this appeared to be a reversion to the 1947 H_1N_1 strain, as a result of which only young people were seriously affected. Vaccines for influenza have been successfully developed but because of the slow antigenic drift of the virus they are usually only useful for short periods of time and may fail totally during a pandemic.

The material produced for a vaccination programme will necessarily have been obtained from a relatively old viral source. After what may sometimes be a relatively short period of time the virus strain circulating in the population will, due to cumulative changes in its surface proteins, have become antigenically distinct from the one originally used to prepare the vaccine. The vaccine then becomes ineffective at protecting from infection and a new one has to be produced. The surprise is that for some viral diseases a vaccination programme works at all! It is usually only when the antigenic surface proteins of the virus have a specific role to play for the virus, e.g. in target cell recognition, that the number of viable mutations is limited and where this limit is small vaccines will be successful.

There are many viral diseases which cannot be controlled by immunological methods and consequently the need for alternative treatments in these cases is paramount. Chemotherapy has just as large a potential for the treatment of viral disorders as it does for other kinds of infections, but the approach which has to be taken to the discovery of such agents is very different because of the close association between viral biochemical processes and those of the cells which they inhabit.

Resistant viruses

Most viruses replicate very quickly; for some species a complete cycle requires less than two hours. Furthermore when viral nucleic acids are replicated, mistakes in their transcription are much more common than for host cell systems because, unlike host polymerases, their replication complexes rarely possess the capacity to edit out mistakes. Consequently the gene sequence may undergo small but cumulative changes which result in the production of proteins with variations in their constituent amino acids. Such changes will have a minimal effect if confined to regions of the protein which are not essential to its function, but may lead to non viable progeny virus if the change does affect its function. In the presence of drug molecules these changes may produce a situation where the effect of the drug is reduced but the viral function is preserved. In these cases the virus is able to continue replicating even though the drug is present and a new strain of the virus emerges which is **resistant** to that agent. Many strains of human pathogenic viruses resistant to particular drugs have been developed under laboratory conditions and others have been identified from clinical isolates. In the situation that two different chemical entities have a similar profile of activity against the same resistant strain, the two compounds are said to be **cross-resistant**. In other words a particular mutation in the target enzyme leads to the same attenuation of activity for both compounds. In the alternative instance that two compounds have reduced activity, but due to a different mutation in the target enzyme for each compound, they are said to be **co-resistant**. In the extreme case a virus may become resistant to many different drugs due to evolving several mutations in the protein sequence of the target enzyme; it is then referred to as **multi-resistant**.

The pattern of resistance of viruses to chemotherapy and the consequences are different for each virus. Often the observed changes are not clinically significant but in some cases useful drugs may be rendered virtually useless within a short space of time. Questions of resistance will be dealt with under the chapters referring to individual viruses.

Viral infections in man

The study of how diseases, including those caused by viruses, are transmitted between individual hosts and disseminated amongst a population is known as **epidemiology**. Intact virus particles have a varying propensity to survive in the 'outside world'. Those which are airborne or surface borne may remain viable for periods of a day or more in suitable conditions and in such cases direct contact may not be necessary for infection of one individual by another. Common colds for example may be transmitted by virus particles adhering to door handles or loose change as well as by direct contact between individuals. Blood-borne viruses, on the other hand, can only be contracted by direct contact of two individuals' body fluids. The common routes of invasion adopted by viruses can be summarized as:

- Respiratory tract mucosae (airborne and surface-borne virus): Influenza; Measles; Varicella zoster (chickenpox); Rhinovirus (colds); Coronavirus (colds); Adenovirus (sore throats); Respiratory syncytial virus.

- Gastro-intestinal tract (faecal-oral transmission): Polio; Rotavirus (viral gastro-enteritis); Marburg disease, Hepatitis A.

- Eye contact: Adenovirus type-8; Herpes simplex (keratitis).

- Genital contact (sexually transmitted): Herpes simplex; Papilloma (warts, cervical cancer); Hepatitis B and C; HIV.

- Small skin abrasions: Herpes Simplex; Papilloma; Vaccinia (smallpox).

- Inoculation of body fluids (transfusions, contaminated needles, etc.): HIV; Hepatitis B and C.

- Inoculation by insect or animal bites: Rabies; Yellow Fever; Other Tropical fevers.

The route of invasion of a virus is not necessarily related to its target organ or tissue; for example polio is taken up through the gastro-intestinal tract and may cause gut irritation but it also has potentially serious effects on the peripheral nerves, resulting in the familiar partial paralysis. There is always a period after primary infection during which no symptoms appear, termed the **incubation period**, which varies from 1–3 days for colds and flu, through months for rabies and hepatitis B, to several years for HIV and some other obscure 'slow' viruses. The first symptoms to appear are highly variable, ranging from superficial rashes (herpes simplex, chickenpox) and mucosal inflammation (colds, measles) to insidious infections of the liver (hepatitis viruses), blood cells (HIV, cytomegalovirus, Epstein–Barr virus) or nervous system (polio, rabies). For viruses such as colds and flu the infections are self-limiting in that they succumb to the host's immune defences and are eliminated from the body relatively quickly. Others such as papilloma and hepatitis B/C lead to chronic conditions which may go on for years. In some cases the virus is able to assume what is called a **latent state**, in which it becomes dormant in certain cells of the body (HIV in T-lymphocytes; herpes simplex in nerve ganglia; Epstein–Barr virus in B-lymphocytes) and becomes reactivated at a much later time for reasons which are not well understood. Sometimes the disease symptoms produced when a latent virus becomes reactivated may be different from those experienced with a primary infection. For example chicken pox, contracted as a mild condition in childhood may re-emerge as shingles in later life; in both cases the causative agent is varicella zoster virus.

For a virus disease to be transmitted from one individual to another, infectious virus particles need to be shed from the host into the environment or directly onto another individual. Shedding of virus may be in aerosols exhaled from the lungs, in saliva, sexual secretions or, in the case of viruses causing rashes, through the skin lesions. Viral shedding usually coincides with the period of maximal viral replication, though not necessarily with the peak of the symptoms.

Main groups of viruses affecting man

Polio, vaccinia (smallpox), measles, yellow fever, rubella, mumps

These viruses, in spite of being from widely different families, are grouped together because it has been possible to develop effective vaccines against them. Prevention of infection is thus straightforward and the need for alternative treatments is minimal. It should appreciated that although vaccination programmes are very effective as a means of prevention of infections the technique is virtually useless for treating non-immunized individuals who have contracted the disease. Chemotherapies for these conditions would therefore have a place in treatment but the impetus for researching them is very small, given also that the economics of a vaccination programme are so favourable.

Colds (rhinovirus, adenovirus, coronavirus, respiratory syncytial virus)

Colds are probably the most frequently occurring viral infections in humans, but the causative agents may be completely unrelated. Furthermore, although rhinoviruses may be responsible for about two thirds of all colds, the enormous serotype variation makes them behave almost as different viruses from the immunological standpoint. Prevention by vaccination is thus out of the question.

Chemotherapy is not easy to countenance either, for most colds are self-limiting infections (i.e. they are eliminated by the host immune system within a short period of time) and the maximum level of viral replication takes place in the first 24 hours or so before the symptoms have maximized. This means that to be effective any treatment would have to be administered almost as soon as the disease is contracted, which for most cases is impracticable. In spite of this, considerable research has been done on the chemotherapy of rhinovirus infections and several compounds have reached clinical trials, details of which are described in Chapter 9.

Influenza

Influenza causes symptoms not unlike colds in the first instance and is, like them, very easily contracted from air- or surface-borne virus. The virus primarily attacks the larynx and trachea producing a general level of malaise which is usually greater than for colds and can for some people be highly debilitating. Serious complications can ensue, especially in the very young and the very old, which may even lead to death. Diagnosis is difficult on a 'one-off' basis but for short periods of time and during epidemics the virus tends to be constant. Short term vaccination programmes for high risk patients have been fairly successful in these circumstances, and are particularly useful in places like old people's homes. Nevertheless the clinical need for symptomatic treatments is considerable and progress towards these is covered in Chapter 9.

Herpes viruses (simplex, varicella zoster, cytomegalovirus, Epstein–Barr)

Diseases caused by viruses of the herpes group have over the past dozen years become the most commercially important for chemotherapy. Many nucleoside analogues have been discovered which selectively inhibit herpes simplex virus, responsible for cold sores, by interfering with its ability to replicate its nucleic acids and one of these, acyclovir, was the first truly selective antiviral drug. Chapter 6 will discuss this field in detail. Varicella zoster virus, which causes chicken pox in early life may re-emerge in middle age when it manifests itself as shingles. This is due to the virus becoming dormant and asymptomatic in the host, a phenomenon known as latency. A satisfactory treatment for human cytomegalovirus (HCMV) which causes serious disease in immunocompromised subjects remains to be discovered. Epstein–Barr virus (EBV) is the aetiologic agent responsible for infectious mononucleosis. It becomes latent in the white blood cells and may be reactivated much later in life with occasionally serious consequences. Most herpes infections are very serious in the immunocompromised population, causing exacerbated, severe symptoms.

HIV and AIDS

Human immunodeficiency virus (HIV) is a retrovirus which was identified in the early 1980s and recognized as the agent which brought about destruction of the immune system as manifest in patients suffering from AIDS (acquired immune deficiency syndrome). Some progress has been made with chemotherapy of HIV but no single agent has proved to be really satisfactory, mainly because of the rapid emergence of resistant viruses. Attempts to produce vaccines have also not yet worked out satisfactorily because of antigenic drift of the immunogenic surface proteins. The problem of treating AIDS remains one of the highest profile problems of modern chemotherapy. Current approaches will be covered in Chapter 7.

Viral hepatitis

Hepatitis, or inflammation of the liver, can be due to many different causes. Among the most serious manifestations are those due to viral infections and there are at least five distinct viruses which target liver cells and result in organ damage. Hepatitis A (HAV) is a picornavirus which is absorbed through the GI tract and causes relatively mild and self-limiting inflammation of the liver. Hepatitis B (HBV) and C (HCV) are blood-borne viruses which may, in a significant proportion of cases, cause long term irreversible liver damage and the possibility of tumours. Some progress has been made with vaccines for HBV but these are not effective once infection has occurred. There are no current therapies for HCV. The full implications of HDV and HEV infections still need to be established.

Papilloma viruses

Human papilloma virus (HPV) is one of a group of viruses which primarily cause warts on skin or endothelial tissue. Many of the warts are benign in that no serious consequences ensue but in some cases the lesions may become malignant and give rise to more serious problems. HPV is particularly implicated in the aetiology of cervical cancer and is thus an important therapeutic target. HPV replicates in tissues which are not accessible to the immune system which means that vaccines are not of any use. For inaccessible lesions in the internal mucosae topical treatments, designed to physically destroy the wart tissue, are not always successful either. Research on the chemotherapy of HPV is in its early days but is receiving increasing attention.

Oncogenic viruses

Several examples of retroviruses have been described which are capable of transforming host cells by inserting so-called oncogenes into the DNA. These cells may then enter an uncontrolled growth phase and lead to the formation of tumours. Such viruses are termed oncogenic viruses. There is very little known at present about what the potential therapeutic targets are for these viruses as a result of which there are no current chemotherapies.

Haemorrhagic fevers

Haemorrhagic fevers are viral diseases which occur mainly in tropical regions. The symptoms are usually serious organ damage resulting in internal and sometimes external bleeding. Specific examples are dengue fever virus and Ebola virus. Their modes of transmission are variable and their symptoms often lead to death but little or no research on their potential chemotherapy has been conducted.

3 | Biological assays for viruses

The drug discovery process

The first step in the discovery of any drug is to select, by some procedure or other, chemical compounds which are thought likely to possess the required activity and then to test them. The sources of compounds which may be available are natural products, existing libraries of compounds or, more often, ideas on paper for which the compounds have to be synthesized from scratch. Although random testing in high throughput assays has its place, the more usual way in which compounds have been selected is in relation to the natural

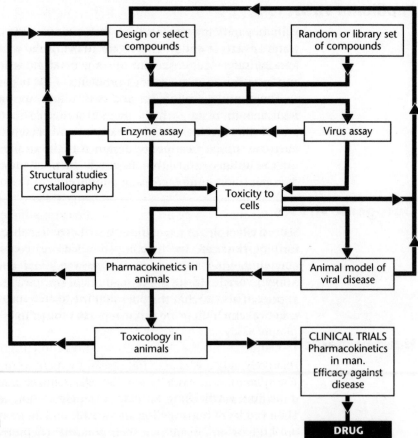

Figure 3.1 Drug discovery for antivirals. The boxes depict all of the stages of the process which will be used if available. It is not unusual for one or more of the primary assay systems, animal model or protein structures to be by-passed. Where high throughput primary assays exist it is increasingly likely that combinatorial libraries of compounds will be used for initial discovery of lead structures.

biochemical processes which drive the target organism. The question of how to decide which compounds to test based on this biochemical knowledge is crucial and is the main theme of Chapter 4.

Discovering active 'lead' compounds is only the first step in the development of a drug. In common with other fields of drug research the process is long, complex and costly. The typical stages are outlined in Figure 3.1.

Cell culture assays

The classical procedure for discovering antibiotics is to grow the pathogenic organism on a suitable culture medium and expose it to varying concentrations of a test compound to see if it remains viable and to determine the level at which the compound inhibits its growth. Given that viruses need to be associated with their host cells in order to replicate, the method of culture has to be different. One way of doing this is to select a susceptible cell line which can be infected by the virus, incubate a free suspension of the cells with infectious virions, and then estimate the amount of progeny virus produced in the culture within a specified period of time. When this is done in the presence of varying concentrations of a test drug, a measure of the inhibitory power of the drug may be deduced. At the same time the effect of the test compound on the uninfected cells may be determined, which gives an indication of the selectivity of the compound as an inhibitor of virus growth compared with its toxic effect on the cells. Ideally, one would want to see a general reduction in the amount of virus produced with no effect on the viability of the host cells.

The amount of virus present in a culture can be measured by an immunological method in which a labelled antigen is added to the culture medium which can recognize the viral particles and become attached to them. Virus can then be estimated by recovering the particles and assaying by means of the label on the antigen. Labels which might be used for this purpose are radioactive isotopes (radio-immunoassay), fluorescent entities or an enzyme linked subunit which can be assayed by a colour change (ELISA) (Figure 3.2).

Virus may also be estimated by the **cytopathic effect** (**cpe**) on cells. A cytopathic effect may be manifest as a change in cell morphology or as cell death, both of which are directly observable. Cell death can be determined by counting the number of viable cells in a culture medium before and after infection with a known volume of virus suspension and comparing with an uninfected control culture. This procedure, when applied to free suspensions of cells, is not very practical as a high throughput assay but the principle has been adapted into the **plaque assay**.

Plaque assays are the most widespread in current antiviral research and are relatively easy and quick, thus enabling many compounds to be tested rapidly. A confluent monolayer of suitable cells is grown in a Petri dish and covered with a medium which does not permit free circulation of virus particles. This can be achieved by using a gelling agent such as agar or by including a virus specific antibody which inactivates free virus but still permits cell to cell transmission.

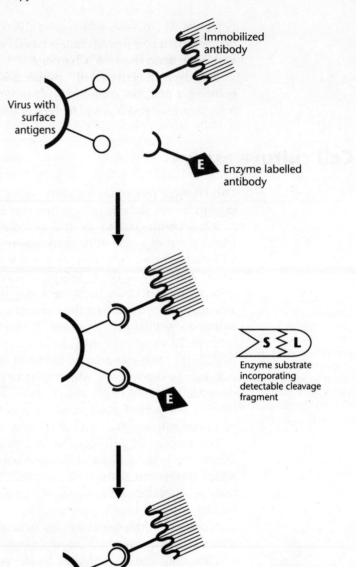

Figure 3.2 Scheme for enzyme linked immunosorbent assay (ELISA). The virus is quantified by the amount of labelled fragment, L, released from the immobilized phase.

Immobilized antibody

Virus with surface antigens

Enzyme labelled antibody

Enzyme substrate incorporating detectable cleavage fragment

Virus–bound enzyme interacts with substrate and releases fragment which can be detected by colorimetric methods

Thus, when virus is included in the medium and infects some of the cells, the progeny virus can only spread to the cells immediately adjoining the original infected cell. After a period of time a circle, or patch, of dead cells will form around the original site of infection and, after fixing and staining of the cell sheet, will appear as a clear patch. This is termed a **plaque**. In practice one plaque forms for each infectious event on the cell sheet; thus a very straight-forward determination of the number of virus particles in the original overlay medium is possible. Likewise for a virus stock of known concentration a fixed number of plaques will form under constant conditions and if a test drug is added, any inhibitory effect will be recorded as a reduction in the number of plaques. The concentration of drug which reduces the number of plaques per plate by 50% is the IC_{50} value for that drug. The system also has the advantage that a preliminary assessment of selectivity can be made since any cytotoxicity of the compound will be revealed by the cell sheets being killed and failing to take the stain. An example of a plaque assay is shown in Figure 3.3

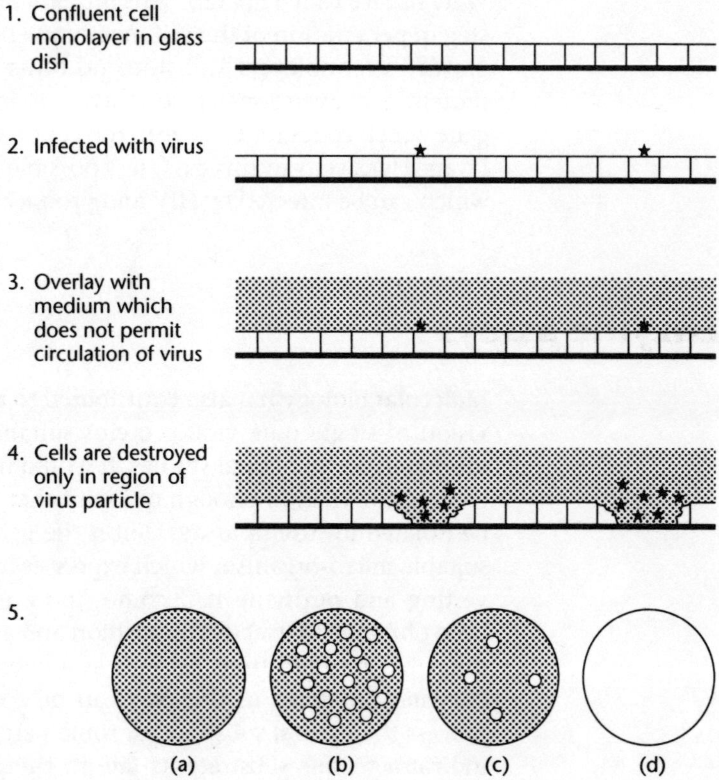

Figure 3.3 Diagrams 1–4: schematic representation of how a plaque assay works. Diagram 5: (a) uninfected cell sheet (cell control); (b) infected cell sheet with plaques (virus control); (c) drug-treated cell sheet with reduced number of plaques from which the percentage inhibition is deduced; (d) toxic compound showing destroyed cell sheet.

1. Confluent cell monolayer in glass dish

2. Infected with virus

3. Overlay with medium which does not permit circulation of virus

4. Cells are destroyed only in region of virus particle

5.

    (a)   (b)   (c)   (d)

Typical cells used for plaque assays are:

- **HeLa:** originally from human cervical carcinoma cells (e.g. rhinoviruses)
- **Vero:** an immortalized monkey kidney cell line (e.g. herpesviruses)
- **MDCK:** immortalized dog kidney cells (e.g. influenza)
- **HeL:** from human embryo lungs (e.g. HCMV)

Plaque assays have enabled huge numbers of compounds to be screened for antiviral activity and have without doubt made a major contribution to drug discovery. There are good plaque assays for herpes viruses, picornaviruses, influenza, yellow fever virus and its relatives and several others of diminishing importance. However, not all viruses can be made to infect cells which are suitable for use in plaque assays; and those which they do infect cannot always be induced to grow in monolayers under conditions where plaques can form. Hepatitis B and C, papilloma and HIV fall into this category and present an awkward hurdle to overcome in screening compounds for antiviral activity.

In such cases it is often possible to genetically modify cells so that they will form plaques. For example, HIV normally infects only T-lymphocytes in the blood. Its specificity is inherent in the mechanism of attachment of the virus particles to the target cells which involves a specific interaction between receptors on the cell surface called CD-4 receptors and a protein component of the virus surface called gp-120. This protein–protein interaction is the essential first step in penetration of the cell membrane by the virus. T-lymphocytes cannot be grown in monolayers and since no other cell types have the essential surface protein a conventional plaque assay is impossible. However, by splicing the gene which codes for CD-4 into the genome of cells which do form stable monolayers, HeLa cells in this case, it is possible to derive a cell culture (HeLa–CD-4+) which can be infected by HIV and produce plaques.

Enzyme assays

Molecular biology has also contributed to the discovery of antivirals by the provision of single pure viral proteins suitable for use in biochemical assays and ultimately for structural studies. It would normally be impossibly labour intensive to grow virus in enough quantity that one of its constituent proteins could be isolated for use in assays. But if the gene encoding it can be inserted into a suitable micro-organism which expresses it in large amounts, the process of harvesting and purifying it becomes more easily manageable. HIV protease has been obtained in exactly this fashion and a new generation of potential antiviral drugs has resulted from using it as a chemotherapeutic target. The activity of enzymes produced in this way can be confirmed by studying their ability to process the natural substrate, or some part of it. By then attaching a fluorescent indicator to this substrate its fate in the presence of a test compound can be assessed and the inhibitory effect of that compound determined. These types of enzyme assay may be used as a primary screen for antiviral compounds and assume great importance when the virus itself cannot be assayed directly in a biological assay. For instance neither hepatitis C nor papilloma virus have yet been successfully grown in readily assayable cell cultures, but sufficient is now known about their genetic make-up that some of their proteins can be characterized, the genes for them cloned and expressed and rapid 'test-tube' assays developed which, for the first time, give a real possibility of discovering drugs for the diseases they cause.

The enzymes which have been most used so far for assays of this type are the polymerases (or transcriptases) which are responsible for processing viral

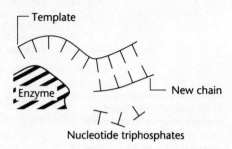

Enzyme attaches new
nucleotides to growing chain

Nucleotide triphosphates

Figure 3.4 Scheme for
measuring polymerase
activity. In a typical assay a
radiolabelled substrate is
used and the degree of
incorporation measured by
scintillation counting. In
the presence of a
polymerase inhibitor the
amount of incorporation
will be reduced in a
concentration-dependent
manner.

If some of the monomers are labelled
they will be incorporated into the
chain by the normal reaction

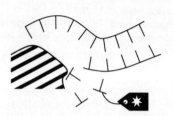

Production of new nucleic acid can
be monitored by detecting
incorporation of the label

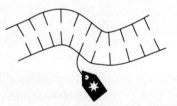

nucleic acids and the proteases which are involved in post-translational protein
processing. Examples of the polymerase assay and protease assay are shown in
Figures 3.4 and 3.5, respectively.

It is important to realize that when screening against an isolated enzyme target it must be separately verified that the test compounds are capable of reach-
ing that target in the living situation. For example the compound must be able
to penetrate cell membranes and must be stable to metabolic degradation. In the
case of most antiviral nucleosides, metabolic activation to their triphosphate
forms is normally essential for activity, a step which has to be performed separately if the target polymerase is being used as an assay.

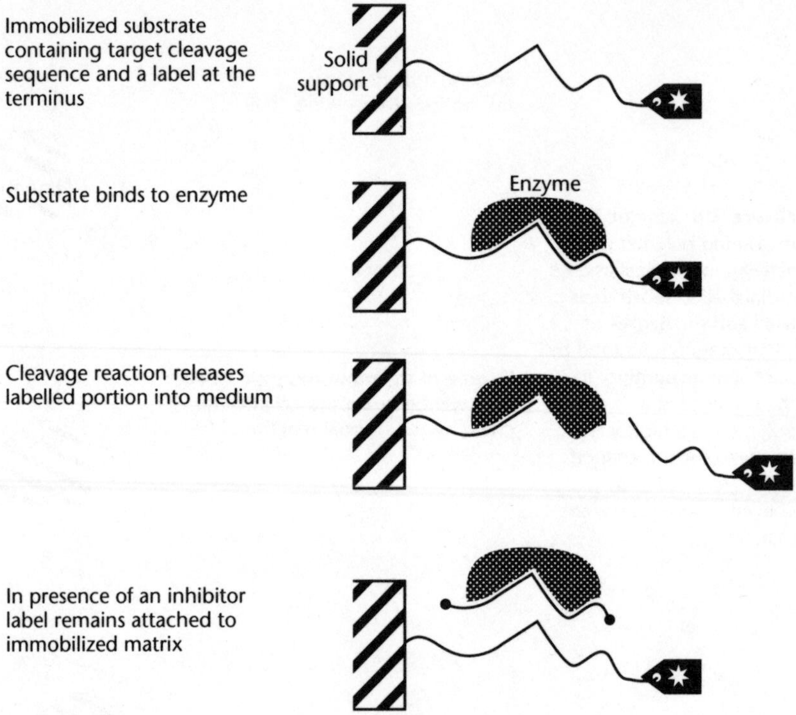

Immobilized substrate
containing target cleavage
sequence and a label at the
terminus

Solid
support

Figure 3.5 Example of an assay
to measure protease activity.
The label may be a radioactive
atom or a fluorescent entity.
Measuring the amount of label in
free solution gives an estimate of
substrate cleavage from which the
effect of the non-cleavable
inhibitor can be inferred.

Substrate binds to enzyme

Enzyme

Cleavage reaction releases
labelled portion into medium

In presence of an inhibitor
label remains attached to
immobilized matrix

In vivo assays

The discovery of a chemical entity which selectively inhibits growth of a virus in
cell culture or one of the viral replication processes in a biochemical assay is no
guarantee that it will be a useful drug. Once it is satisfactorily established that an
active compound is non-toxic to normal cells it is essential to know something
about how the compound is handled in a typical host organism. Questions
requiring answers are:

- Is the compound orally absorbed?
- Does it reach the required site of action at sufficient concentration and for
  long enough to have a beneficial effect?
- Is the compound stable?
- Does it have unpredicted toxic effects?

Antivirals are no different from drugs for any other purpose in these respects
and the necessary pharmacokinetic, metabolic and toxicology studies are integral parts of a drug development programme.

A further useful indicator of the efficacy of a candidate drug is a suitable
animal model. The model should preferably be in a small animal since only relatively small amounts of compound are then required to demonstrate efficacy.
Some animal models commonly used in anti-viral research are given in Table 3.1.

Table 3.1 Animal models of viral disease

Virus	Susceptible animal	Comment
HSV-1	Mouse	Zosteriform model
HSV-1	Guinea-pig	
MCMV	Mouse	Surrogate for HCMV
Influenza	Mouse	Lung infection
Influenza	Ferret	Parameters similar to human
VZV	Chimpanzee	Rarely used
HIV	SCID mouse	Immunocompromised animal
HBV	SCID mouse	
HBV	Woodchuck	Inconvenient in use
Coxsackie	Mouse	Surrogate for human picornaviruses

Herpes simplex-1 infects mice and produces symptoms not unlike those of shingles in humans (the zosteriform model). The condition is relatively easy to monitor as its effects on the skin of the animal can be observed externally. It has been used extensively in the discovery of anti-herpetic drugs to assess which compounds are the most likely to be effective in humans. The situation is not so straightforward with other viruses of the herpes group. There is a virus closely related to human cytomegalovirus which infects mice (murine CMV or MCMV) and produces effects on the internal organs of the animal, such as the spleen, which can be measured both by physical examination and by titrating the virus contained in the organs. The MCMV model, which can be described as a **surrogate animal model**, has proved useful for selecting anti-HCMV compounds, particularly nucleosides, for further study. For varicella zoster there is no small animal model (unless one includes chimpanzees, which are very rarely used), which means that in cases where the test drugs are only active against this virus clinical trials have to be conducted on the basis of finding active compounds which have a suitable pharmacokinetic and safety profile.

Influenza virus affects mice to produce a lung condition which can be monitored by measuring virus titres in the lungs of sacrificed animals. It also infects ferrets to produce a range of symptoms similar to those in humans. Both models are used for assessing the efficacy of anti-influenza compounds.

Sometimes it is possible to breed animals with altered genes which make them susceptible to infections which would not normally be observable. In particular a genetically immunocompromized mouse, the severe combined immunodeficient, or SCID, mouse is especially useful. These animals will accept HIV-infected T-lymphocytes or xenografts of HBV infected human tumour cells and maintain their viability for long enough to test potential drug molecules *in vivo*. Even so, there remains a significant group of viruses, including rhinovirus and hepatitis C, for which there is presently no adequate animal model.

In some cases a virus of the same genus as the target one and having some common morphological and genetic features will infect a small animal. If both viruses are susceptible to the compound(s) under investigation the model may be used to assess activity *in vivo* and to form some idea of how the compound is handled in an intact host. Such cases are referred to as **surrogate viruses**. An example is the use of yellow fever virus as a surrogate for Hepatitis C.

4 Biochemical targets for chemotherapy

Viral replication cycles

After describing the main characteristics of viruses in previous chapters, this chapter focuses on the biochemical role that each of the component parts plays in the viral replication cycle and on the consequent opportunities for therapeutic intervention. There are general similarities between viruses in the way they replicate but there are also unique features which can be exploited as drug design targets.

The general pattern of viral replication cycles, whatever the detail differences, comprises events which are mediated by the various viral components. The driving force behind the existence of a virus, in common with every other life form, is to ensure the perpetuation of its nuclear material, be it RNA or DNA. It must therefore contain the means of replicating its nucleic acids when the environment is appropriate and the means of ensuring that they are preserved at other times by being safely packaged into suitable particles. Along with this, the virus must include a means by which its material can access a suitable environment for replication and a mechanism for creating its protective packages, or particles.

Six discrete stages of the replication cycle are important which are summarized as:

- **attachment** to the host cell
- **penetration** of the host cell membrane
- **uncoating** of the virus to release its core components
- **replication** of the core nucleic acids and translation of the genome
- **maturation** or reassembly of progeny virus particles
- **release** of progeny virus into the environment

The viral replication cycle is also outlined in Figure 4.1

Attachment to the host cell

All viruses contain protein material on their surface, whether as an integral part of the capsid structure or in the form of spikes embedded in the viral envelope. One of the main functions of the surface proteins is to recognize specific

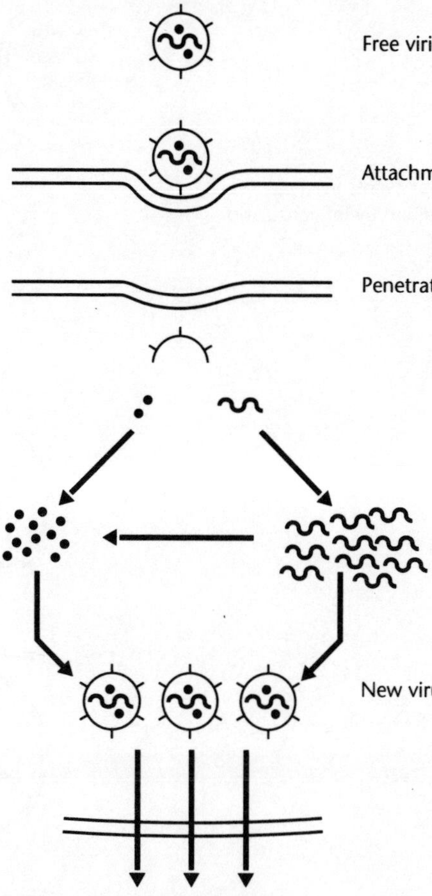

Figure 4.1 General representation of a viral replication cycle. The detailed mechanism for each stage may have considerable variation.

Free virion

Attachment to cell membrane

Penetration and uncoating

Replication of protein and nucleic acids

New virus

Release from cell

receptor proteins on the surface of the target host cells. This recognition event is similar to any other protein–protein interaction in that it occurs through a stereospecific network of hydrogen bonds and lipophilic associations.

An example is the association between haemagglutinin (HA), one of the surface proteins of influenza, and the termini of cell surface glycoproteins called gangliosides. Each molecule of HA contains a conserved pocket which is capable of recognising the terminal glycoside residue of the cell surface gangliosides and leads directly to the virus particle adhering to the cell (Figure 4.2). Chapter 9 deals with this topic in more detail.

A similar interaction occurs during the attachment of HIV virus particles to T-lymphocytes whose membranes contain a protein known as CD-4. The virus has a surface protein called gp-120 which is specifically capable of recognising CD-4 and the resulting complex attaches the virus particle to the cell. The reason why HIV only primarily infects T-lymphocytes is because other cells do not possess a compatible receptor. In other cases the virus–receptor interaction is specific for the cells of a particular species as well as a particular cell population. It is for this reason that poliovirus only attacks humans.

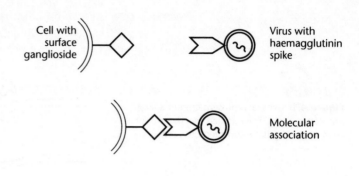

Figure 4.2 Sequence of events involved in cell surface receptor recognition by influenza and absorption into the cell.

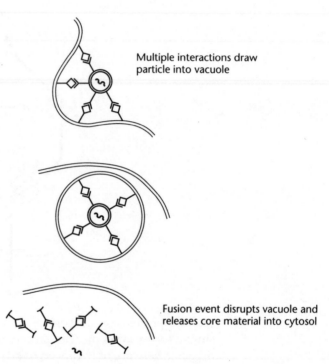

Because all viruses target their host cells in this manner, the opportunity for interfering with virus–target cell interactions has considerable attractions since the chemical agent employed does not itself need to enter the cell and this could have beneficial consequences from the point of view of selectivity of action. However, for such an approach to be totally effective, all virus–cell interactions would need to be neutralized and since the numbers of such interactions per virus particle can be very high (several hundred for influenza HA) the affinity of the drug molecule for the viral receptor would need to be very high too. There is also the problem of timing of therapy. Often a viral infection does not reveal symptoms until it is well established in the host. A therapy which relies for its effect on the initial process of infection might, in such circumstances, not be very effective and would need to be used as a preventative (prophylactic) treatment. Producing antivirals by this means has not yet been successful.

Nevertheless, vaccines work by stimulating the immune system to act in precisely this fashion and so the possibility of better progress in the future, possibly with a biological agent, should not be overlooked.

Penetration of the cell membrane and uncoating of the virus

The next process which must be accomplished by the virus, once attached to the exterior cell surface, is to insert its genetic material and biochemically active proteins into the cell interior, which means that the lipid bilayer surrounding the cell has to be penetrated. In the case of enveloped viruses the consequence of association with cell surface receptors is to draw the cell membrane around the virus particle and enclose it within a cytosolic vacuole (a process called **pinocytosis**), similar to the process by which cells ingest other materials essential to their function. The pH inside such vacuoles begins to fall which stimulates the viral surface protein (HA in the case of influenza) to change conformation and trigger the fusion of the viral envelope with the membrane of the cell. The respective membranes become fused into a continuous lipid bilayer, as a result of which the core material is released into the cytosol and the matrix protein is stripped off. Figures 4.3 and 4.4 represent the penetration sequence at the cell surface, and the endocytic penetration sequence.

Cell membrane

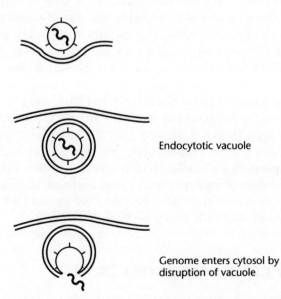

Figure 4.3
Penetration event involving endocytosis before membrane fusion. The membrane fusion event is triggered by a pH change inside the vacuole.

Endocytotic vacuole

Genome enters cytosol by disruption of vacuole

With non-enveloped viruses the viral coat protein interacts with cell membrane-bound proteins to open up a 'pore', through which the viral core material can

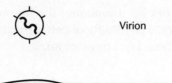

Virion

Cell membrane

Cytosol

Figure 4.4 Typical penetration/uncoating sequence for a small non-enveloped virus.

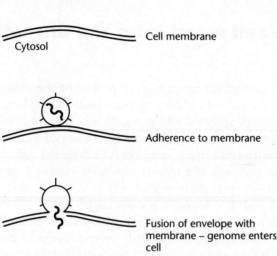

Adherence to membrane

Fusion of envelope with membrane – genome enters cell

pass into the interior of the cell. For simple viruses, such as picornaviruses, the injection of the core material into the cell in this manner involves concurrent uncoating and immediate exposure of the cytosol to the functional components of the virus which may then begin replicating.

The potential exists for chemical agents to bind to the coat protein in such a manner that the fusion and/or uncoating process is interfered with. In such a case the viral cores would remain in an inactive state and replication could not commence. There are several examples of this type of antiviral agent which have been, or are still being, investigated, including flavanoids and chalcones. One series of oxazolidines, active against rhinoviruses, is known from structural studies to act by binding strongly to the viral capsids, thereby stabilizing them to uncoating. The main disadvantages of the compounds are that they are only active against a limited range of serotypes and have very poor efficacy in clinical trials for reasons which have not been satisfactorily explained. Nevertheless, the strategy has been shown to be very effective for other viruses. The lipophilic amines, amantadine and rimantadine, are proven treatments for influenza A infections operating by binding to ion channels in the matrix protein of the virus and preventing secondary uncoating. There are also some large ring heterocyclic compounds recently described as active against HIV which are reported to work by interfering with the virus uncoating process.

Replication of viral core components

Once the viral proteins and nucleic acids are inside the cell they can begin functioning. Two basic biochemical processes are involved: synthesis of new viral proteins (translation of the genomes) and new viral nucleic acids, (nucleic acid transcription).

Translation of the genome

Translation of the viral genome into its constituent proteins is one main biochemical event which occurs during viral replication. The virus invariably takes over the cellular apparatus for this purpose, utilizing its own messenger RNA to make proteins in the ribosomes of the cell. Many viruses also produce proteins which concurrently suppress the formation of natural cellular proteins while its own material is being replicated. Viral protein synthesis can be subdivided into **early protein synthesis** and **late protein synthesis**. Early synthesis is concerned with the rapid production of those functional proteins which are necessary to expedite viral replication, such as polymerases, whilst late synthesis is more concerned with the production of the structural proteins which are essential for assembling new virions. The relative rate of synthesis of the various proteins, especially in complex viruses, is under the control of virally encoded regulatory factors, either carried by the virus particles or produced during the early phase, but in either case the basic mechanism is the same and requires suitable messenger RNA.

Positive strand RNA viruses (e.g. rhinovirus) can use the parent RNA as mRNA. It acts by acquiring a terminal sequence from the host after which translation into proteins is immediately possible. Negative strand viruses (e.g. influenza) use the positive strand RNA from primary transcription for the same purpose.

DNA viruses (e.g. herpes viruses) make their mRNA by using their template DNA in conjunction with host DNA-directed RNA polymerase II. The process is little different from that utilized by the host cell.

Retroviruses (e.g. HIV) do not use their positive strand RNA directly for translation but instead it is the proviral DNA, inserted into the host chromosomes, from which the viral proteins are derived using the intact host cell mechanisms.

Since host cell processes are directly involved in viral protein synthesis the opportunities for intervention by therapeutic agents are grossly limited. There is, however, the possibility that post-translational protein processing can be a target for chemotherapy. Unlike host mRNAs, which code directly for functional proteins, most viral genomes are polycistronic, which means that all the genetic information either is or becomes encoded in one piece of mRNA. When this is translated, a single polyprotein is produced which is biochemically inert. To convert it into functional proteins and structural proteins it has to be split into its active components by cleavage at the appropriate positions. All viruses which have a polycistronic genome therefore contain a virus specific and **virally coded protease** which is capable of performing the cleavages in the correct manner. Many of these viral proteases have been characterized during recent years and their validity as chemotherapeutic targets has been demonstrated, especially for HIV, where many inhibitors have been discovered.

Nucleic acid transcription

The general mechanism of nucleic acid replication is well understood (see Chapter 5). The way in which these processes are handled in viruses varies slightly depending upon whether the viral genome is encoded in RNA or DNA. For positive strand RNA viruses the virally encoded RNA polymerase transcribes the RNA into a complementary negative strand which remains associated as double stranded RNA. This is then copied into more positive strands to amplify transcription by repeating the cycle, ultimately for incorporation into new virus particles. For negative strand RNA viruses the RNA polymerase makes a complementary positive strand and the resulting double strand RNA is used to produce new negative strands for incorporation into new virions.

Whether the original viral genome is in the form of a positive or a negative strand, the **RNA-directed RNA polymerase** which controls these events is specific to the virus and has no counterpart in mammalian biochemistry. It is therefore potentially a good target for chemotherapy.

Human DNA viruses, such as herpes viruses, already possess a double stranded genome which is replicated as a double strand by a virally encoded **DNA-directed DNA polymerase** similar to the mammalian DNA processing enzymes. A number of nucleoside drugs work by means of their respective triphosphates inhibiting the viral polymerase or acting as chain terminating substrates. As discussed later, subtle differences in the phosphorylation of the parent nucleosides may make an

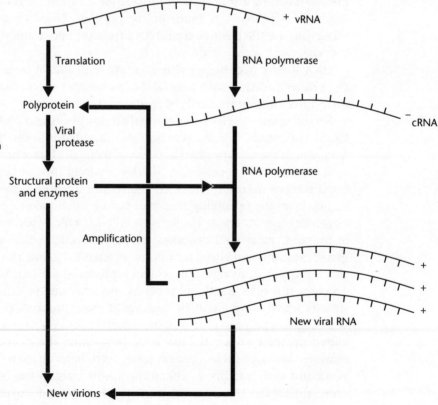

Figure 4.5 Replication scheme for a positive strand RNA virus with a single strand genome. Complementary (negative) RNA is required as a template for new viral RNA synthesis.

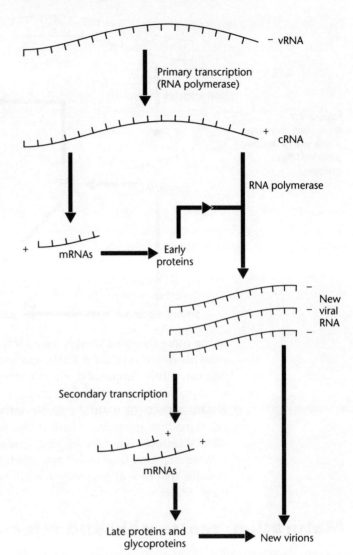

Figure 4.6 Replication scheme for a negative strand RNA virus. The viral RNA must first be transcribed into a positive complementary strand in order to provide mRNA for translation.

important contribution to selectivity. Replication schemes for DNA and RNA genomes are shown in Figures 4.5, 4.6, and 4.7

Retroviruses, containing a single positive strand of RNA, replicate by a rather different mechanism. First of all the genome is transcribed into complementary DNA which then forms **hybrid nucleic acid**, where one chain of the double strand is RNA and the other is DNA. The RNA is then digested by a **ribonuclease** and replaced by a positive DNA strand and the resulting duplex is then integrated into the host genome where it is known as **proviral DNA**. Towards the end of the replication cycle molecules of the original positive RNA accumulate ready for incorporation into new virus. In HIV most of these events are catalysed by the enzyme **reverse transcriptase** (**RT**) which is unique to the virus. A large number of chemical agents are now known which inhibit the function of RT.

Although nucleic acid synthesis is primarily catalysed by polymerases, many

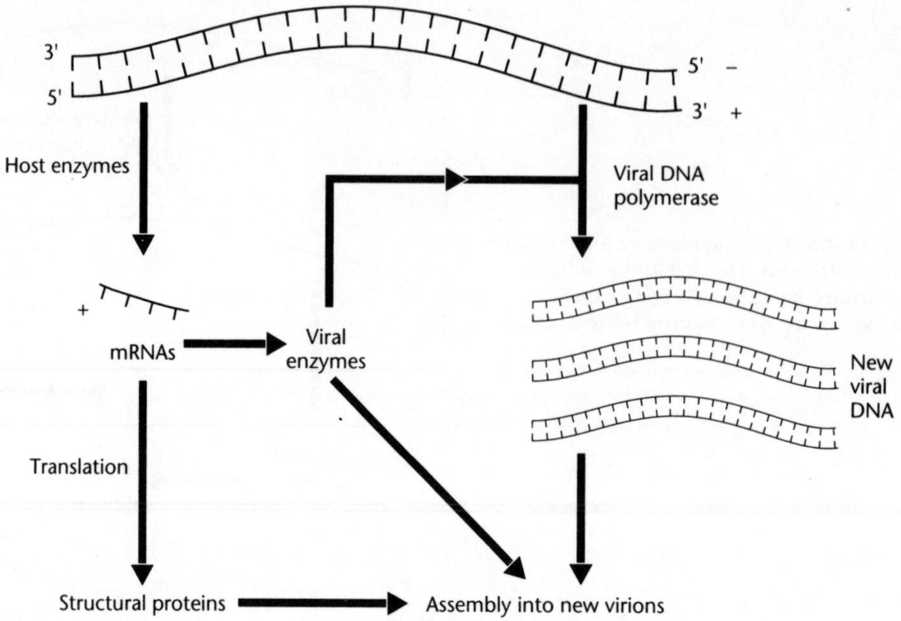

Figure 4.7
Replication sheme for a virus with a double-stranded DNA genome.

of the more complex viruses code for a number of **regulatory factors**. These are proteins which interact with the cellular processes in such a way that host functions are either suppressed or accelerated according to the needs of the virus. For example, when HCMV infects its host one of the general effects is to stimulate certain aspects of natural cellular activity, probably through the endogenous cell signalling apparatus. Much is also now becoming understood about the regulatory factors coded for by HIV, one of which (TAT) has been shown to be a valid therapeutic target with active inhibitors. Clearly the field is young and it is possible that new approaches to antiviral chemotherapy may soon emerge from such studies.

Maturation, reassembly and release of virus particles

As the host cell supply of raw materials becomes exhausted, large amounts of viral material accumulate in the cytoplasm and it is at this stage that the role of the structural proteins becomes paramount. The structural proteins of most viruses possess the ability to self-assemble once they are formed.

Small non-enveloped viruses such as picornaviruses do this automatically as viral components are synthesized. The structural proteins form initially into capsomeres which then associate into capsids and enclose the genome along with such functional proteins as are necessary to initiate viral replication again in a new cell. Eventually the cell becomes totally exhausted, it dies, breaks open and the progeny virus particles are released into the extracellular environment.

Enveloped viruses behave differently. Here the structural, or matrix, proteins migrate to the inside of the cell membrane where they control the sequential assembly of the virus. Core material becomes associated with the matrix protein and surface components are inserted into the neighbouring region of the cell membrane. Once all is in place the matrix protein induces the membrane to gather round the viral centre and form a protuberance on the cell surface which then seals itself from the interior of the cell and is released into the extracellular environment, the process being known as **budding**. Enveloped viruses can therefore be released continuously from cells which are still living.

In many ways the release of virus from its host cell is the reverse of attachment and entry and it might therefore be expected that compounds which can bind to the various assembly proteins could interfere with the process and inhibit virus production. In practice this is difficult to observe since such compounds would be more likely to have their effect on the early stages of the replication cycle and the opportunity to inhibit reassembly would not arise. However there is evidence that some antivirals discovered by screening may act late in the replication cycle. Additionally there is a very good example of a rationally designed compound which inhibits budding of influenza. Most of the protein spikes on the surface of influenza consist of haemagglutinin but a small minority are a different protein which has the ability to cleave the terminal residues from cell surface gangliosides and is called **neuraminidase (NA)**. It is probable that this enzyme activity is responsible for the final step of detachment of new virus from its cell and for maintaining its infective properties. Recently, powerful inhibitors of NA have been discovered which have demonstrable efficacy in animal models of influenza.

Interferon

Interferons are endogenous proteins produced by cells in response to certain external stimuli. Their properties are such that they can be considered to be antiviral agents, even though they do not fit comfortably into any conventional class of chemotherapeutic agent. The first interferon was reported in 1957 by Isaacs and Lindeman as a soluble protein of molecular weight about 20,000 kDa produced by leucocytes in response to infection by viruses. Subsequently, it has been shown that there is more than one type of interferon depending on its cellular source, termed α-IFN (from leucocytes), β-IFN (from fibroblasts) and γ-IFN (from T-lymphocytes). They are active against any virus but are unique to the originating species, which means that only interferons from a human source can be used for treating human infections.

The modes of action of interferons are diverse but one of the most important effects is that the enzyme, ribonuclease, is stimulated which results in the rapid digestion of RNA in the cell. Viral replication is crucially dependent upon the role of RNA, especially for viruses with an RNA genome; hence removing it results in inhibition, or slowing, of virus growth.

Until recently supplies of IFNs were restricted which limited the number of investigations which could be carried out. But since the 1980s cultured and genetically engineered interferons have become available, and have been used in a number of studies against viral diseases in man, including the common cold. So far the only real promise appears to be for the treatment of viral hepatitis where clear clinical benefits have been demonstrated.

5 | Nucleoside analogues as antiviral agents

Introduction

Nucleic acids are essential components of all living organisms; they carry the hereditary information to bring about faithful reproduction and contain the blueprint which directs the synthesis of all of the organism's proteins. As discussed in Chapter 2, viruses consist of little more than a **nucleic acid** core, which may be of either DNA or RNA, packaged within some kind of coat. Thus, targeting this nucleic acid and its functions have, unsurprisingly, received the most attention as therapeutic targets. When an invading virus takes over a host cell's biosynthetic machinery, the replication of this genome is fundamental to viral reproduction. However small the viral genome, it almost always encodes one or more enzymes which have vital functions in the replication of the viral nucleic acids. Such **viral polymerases** have been the most widely studied targets for antiviral chemotherapy, given that by selectively blocking the production of viral nucleic acids, viral replication is shut down. Much activity has been centred on nucleoside analogues which mimic the natural substrates of these enzymes and provide good starting points for chemical manipulation. Nucleoside analogues, or derivatives, *vide infra*, which act against these enzymes, need to be highly selective for the viral enzymes without interfering with the action of similar host functions. Much of this chapter focuses on the inhibition of viral nucleic acid polymerases by nucleoside analogues, which is the single most important chemotherapeutic interaction to date, with brief descriptions of other nucleic acid based approaches.

Nucleic acid structure

The basic structure of nucleic acids comprises a chain of **D-Ribose** sugars interlinked by **phophodiester** bridges between the 3'- and 5'-hydroxyls of adjacent sugars. The sugar is D-Ribose in **Ribonucleic Acid** (RNA) and **2-Deoxyribose** in **Deoxyribonucleic Acid** (DNA). Variation occurs in the **bases** attached to β-face of the 1'-position of the ribosyl unit; which may be either one of the **purines—adenine** or **guanine**; or a **pyrimidine—cytosine**, **uracil** (RNA only) or **thymine** (DNA only) will be present. Examples of natural nucleosides are shown in Figure 5.1. In living systems the nucleic acid normally exists in duplex form, making the characteristic double helix through specific base pairing. Biosynthetically, these macromolecules are assembled by attaching each constituent **nucleoside**

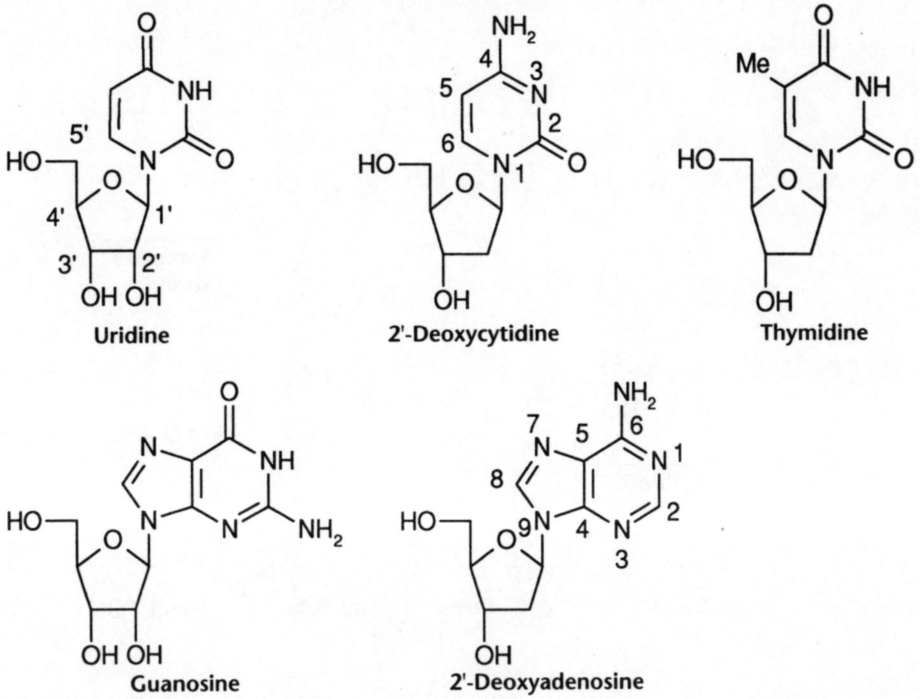

Figure 5.1 Examples of natural nucleosides. The numbering conventions are also shown.

Uridine

2'-Deoxycytidine

Thymidine

Guanosine

2'-Deoxyadenosine

monophosphate derivative (or **nucleotide**) to the growing chain. Fidelity in the replication cycle is ensured by the molecular recognition, by hydrogen bonding, of complementary G–C, C–G, A–T and T–A pairs in the synthesis of DNA. When RNA is formed from a DNA template, uridine replaces thymidine in the transcribed RNA molecule. Nucleoside nomenclature* is given in Table 5.1, and DNA polymerization is shown in Figure 5.2

Table 5.1 Nucleoside Nomenclature

Base		Nucleoside	Nucleotide (5'-Monophosphate)
Adenine	(A)	Adenosine*	Adenylate* (AMP)
Guanine	(G)	Guanosine*	Guanylate* (GMP)
Cytosine	(C)	Cytidine*	Cytidylate* (CMP)
Thymine	(T)	Thymidine (DNA)	Thymidylate (TMP)
Uracil	(U)	Uridine (RNA)	Uridylate (UMP)

*All have 2'-deoxyriboside equivalents.
Thymidine has no riboside equivalent. Uridine has no 2'-deoxy equivalent.

*Throughout the book, nucleoside structures are drawn using a conventional flat representation, whereby bonds from the sugar directed upwards are above the plane of the sugar and those downwards below the plane. Thus thymidine is represented using the conventional structure which implies the stereochemistry shown in the hashed/wedged representation alongside.

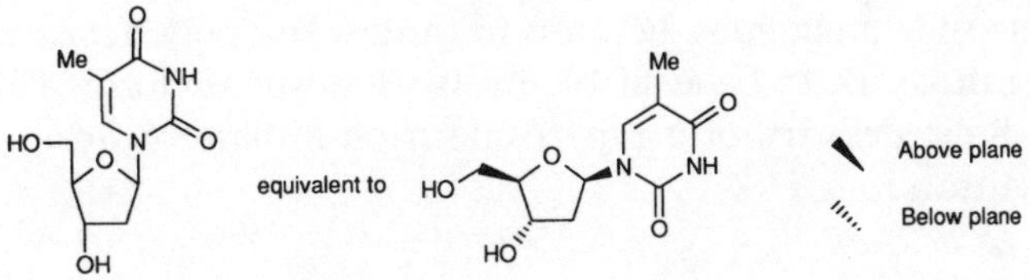

equivalent to

Above plane

Below plane

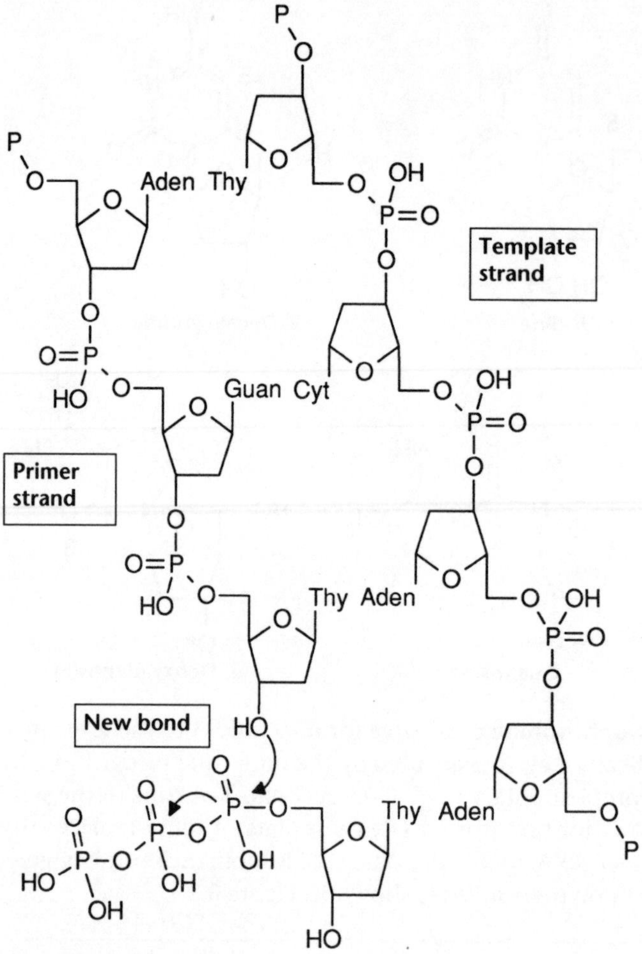

Figure 5.2 DNA polymerization. This short DNA duplex shows the sugar–phosphate backbone and the key DNA base pairings. The incorporation of the next thymidine in the sequence, by reaction of the primer strand's 3'-hydroxy with thymidine triphosphate, is shown; the key event here is recognition by complementary adenine residue in the template DNA strand.

Phosphorylation of nucleoside analogues

When nucleic acids are built up, the monomeric substrates for the polymerase enzyme are the **nucleoside triphosphate** esters, which are produced by reversible sequential phosphorylation, under the control of **kinase enzymes**. For example, thymidine is phosphorylated (Figure 5.3) by a specific **thymidine kinase** to its **monophosphate** which is in turn taken up to the **di-** and **triphosphate** level by **thymidylate kinase**. The specific requirement of a **triphosphate** for nucleic acid elongation has important implications in the design of potential anti-viral nucleoside analogues, because to inhibit the polymerase it is essential for the drug molecule to be available in its triphosphate form. This could be achieved by direct delivery of a nucleoside triphosphate or using *in situ* activation of precursors.

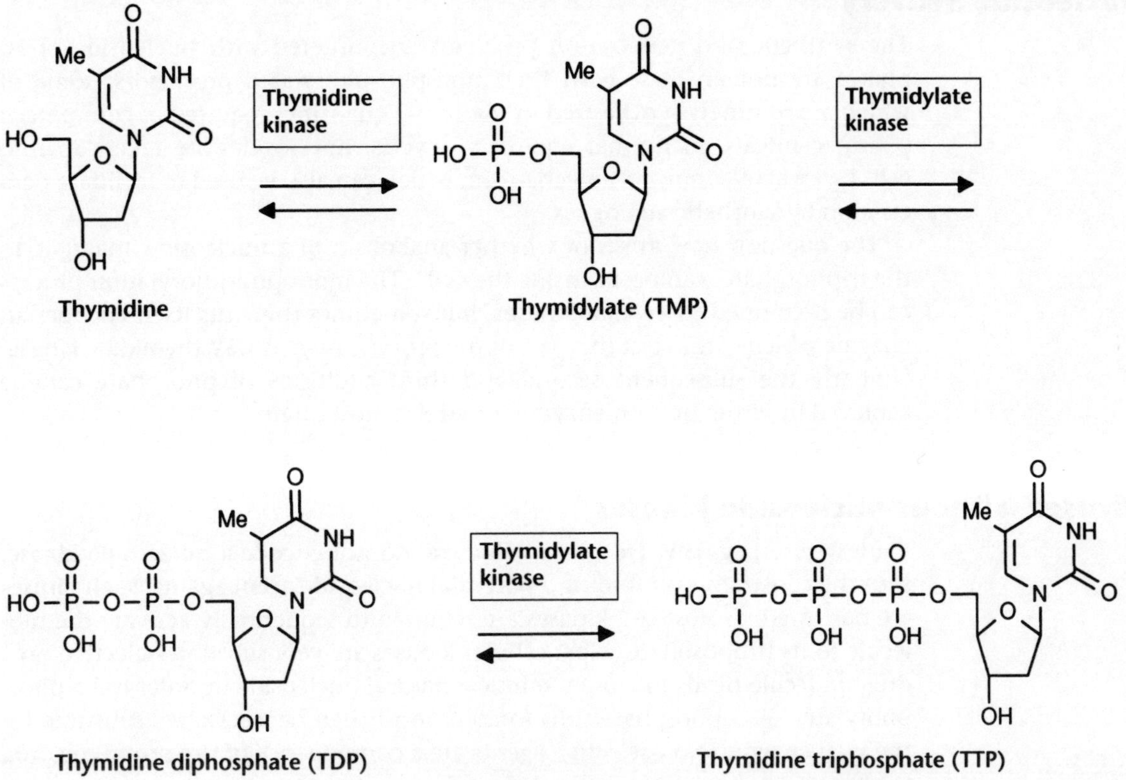

Figure 5.3 Thymidine phosphorylation.

Nucleoside phosphate delivery?

It is possible to chemically synthesize nucleoside triphosphates in the laboratory, although this is a relatively difficult operation and impractical on other than a small scale. Even if such molecules were freely available they would still be useless as drugs, as the molecules are too polar to pass through cell membranes and thus cannot reach the site of viral replication. Nucleoside mono- and diphosphates do not penetrate cell membranes either, but in the former case a potentially attractive option is to use a **prodrug** form which, on intracellular cleavage, liberates the desired monophosphate, a substrate for enzymatic conversion to the triphosphate. This area has been relatively well studied and has enabled the intracellular production of nucleoside triphosphates. This can work favourably when the triphosphate is a selective inhibitor of the viral polymerase over host enzymes, especially when the virus does not code for a kinase enzyme or the first phosphorylation is slow or not achievable. Should the triphosphate inhibition be poorly selective, however, such prodrugs are a good way of producing cytotoxic compounds, indeed the advantages due to virally encoded kinase enzymes are now lost.

Nucleoside delivery

The synthetic and penetration problems encountered with nucleoside phosphates are not an issue with their non-phosphorylated precursors, some of which are routinely synthesized in the tonne quantities required of commercial pharmaceuticals. In normal replication cycles, nucleosides are taken up into cells by an active transport mechanism, which can also be used to facilitate penetration by synthetic analogues.

The question now arises of whether anabolism of a nucleoside analogue to the triphosphate, can occur inside the cells. The monophosphorylation process can be performed by **cellular kinases**, but sometimes the virus itself encodes an enzyme which carries out this part of the process, such as **HSV thymidine kinase**. Similarly the subsequent second and third additions of phosphate can be achieved by either or both enzymes of viral or host origin.

Viruses without nucleoside kinases

Many viruses, eg. **HIV**, **HBV** and **influenza**, do not encode a nucleoside kinase, thus the constraints on finding a potential nucleoside agent against such viruses are narrowed, as host cell kinases are required to sequentially activate the molecule to its triphosphate. Most cellular kinases are very substrate-selective, so a drug molecule needs to closely mimic a natural nucleoside in order to be phosphorylated. A triphosphate thus formed, must then be a selective substrate for the viral enzyme. Non-selective agents are a consequence of the prototype drug triphosphate also being an inhibitor of, or a **chain terminating** substrate for, host nucleic acid polymerases. Compounds which inhibit cellular as well as viral enzymes, will be revealed as cytotoxic in the *in vitro* screening; this may well be cell-type dependent, given differing activity levels of specific kinases in various cells. This may also affect virus targeting, as many infect specific cell types. Numerous potential agents screened against **kinase-deficient** viruses have fallen down due to their phosphorylation properties; either they have proved inactive because they cannot be phosphorylated *in situ*, or they were phosphorylated, but the triphosphate was then also an inhibitor of the host polymerases. Synthetic triphosphates of molecules which are inactive in cellular assays have sometimes been shown to be selective inhibitors of viral polymerases; the enantiomers of carbovir provide a good example as both triphosphates inhibit HIV-RT, but only the naturally-configured enantiomer is active in cell cultures due to selective phosphorylation. The activation of nucleoside analogues that require host enzymes for phosphorylation is shown in Figure 5.4.

Viruses with nucleoside kinases

Viruses which have a larger genome, notably members of the herpes group, sometimes encode their own nucleoside phosphorylating enzymes. This is a very important factor in their chemotherapy, as now there is a potential means of activating nucleosides **selectively** in infected cells. The initial phosphorylation is usually the most difficult to achieve, monophosphates being generally

Infected or uninfected cells
Compound *is not* a substrate for cellular kinases

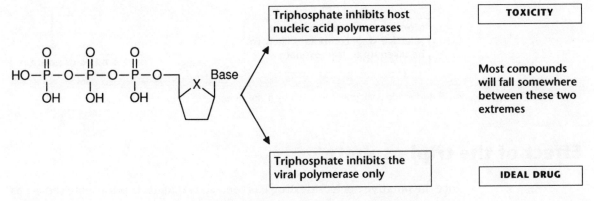

Figure 5.4 Activation of nucleoside analogues that require host enzymes for phosphorylation.

more readily anabolized to the triphosphate. It is feasible therefore that if the drug were a substrate for the viral kinase, but not the cellular kinases, it might be activated only in virally infected cells. This is a major factor in the success of acyclovir (ACV) as a chemotherapeutic agent, coupled with the high selectivity of ACV triphosphate for the inhibition of the viral DNA polymerase over host enzymes. Indeed, if the activation is specific, then the selectivity of the triphosphate for the viral polymerase is not so important, because only infected cells will contain the triphosphate; this is the case with the anti-VZV agent zonavir. It is also possible for nucleosides to be phosphorylated by enzymes other than specific nucleoside kinases; the CMV gene product UL-97 is thought to be responsible for the phosphorylation of ganciclovir. Phosphorylation of nucleoside analogues mediated by a viral kinase is illustrated in Figure 5.5

Uninfected cells

Infected cells

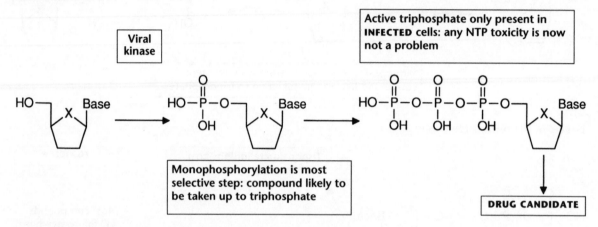

Figure 5.5 Phosphorylation of nucleoside analogues mediated by a viral kinase.

Effect of the triphosphate

Once the putative nucleoside drug has been activated, its triphosphate can act as either or both, a competitive inhibitor of the viral enzyme or a chain terminator of the nucleic acid processing. The nucleoside might also exert an inhibitory effect when it is incorporated as a link in the chain, even when it is not a mandatory chain terminator, such as an arabinoside or a 2'-deoxy riboside.

Design of nucleoside analogues as antivirals

Analogues of the natural substrates for nucleic acid processing enzymes represent a good source of potential antivirals. As the preceding sections suggest, a vital element of the structure of a potential anti-viral nucleoside is the hydroxymethyl functionality which must be converted to a triphosphate ester. There are two further elements which are sources of variations in analogues; these are the sugar component, of which the hydroxymethyl group is a part, and the heterocyclic base attached to this sugar. The interplay of these elements has provided enormous scope for exploitation by medicinal chemists.

A nucleoside triphosphate has two features which are fundamentally important to its molecular recognition: the triphosphate ester itself and the heterocyclic base. The triphosphate, a charged, polyoxygenated 'arm', is electrostatically bound via chelating metal ions to the enzyme structure, whilst the base is hydrogen bonded to its complementary nucleotide on the DNA template. The same features prevail in the recognition of substrate molecules by the intermediate phosphorylating enzymes; even though the base will not be paired to a complementary nucleotide, its features are still recognized. In this context, the sugar seems merely to act as a spacer (Figure 5.6) or a scaffold to present these elements in the correct orientation, which is reflected in the numerous variations which have been made whilst still retaining activity.

Figure 5.6 Generalized structure of a nucleoside analogue.

Base recognition

The recognition due to specific hydrogen bonds (Figure 5.7) between the nucleic acid bases is arguably the most important interaction of its kind, fundamental in the reproduction of the genomic nucleic acid material of all living organisms. So, exploiting such an interaction targeting viral nucleic acid polymerases is fundamentally important in the design of inhibitors.

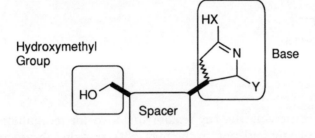

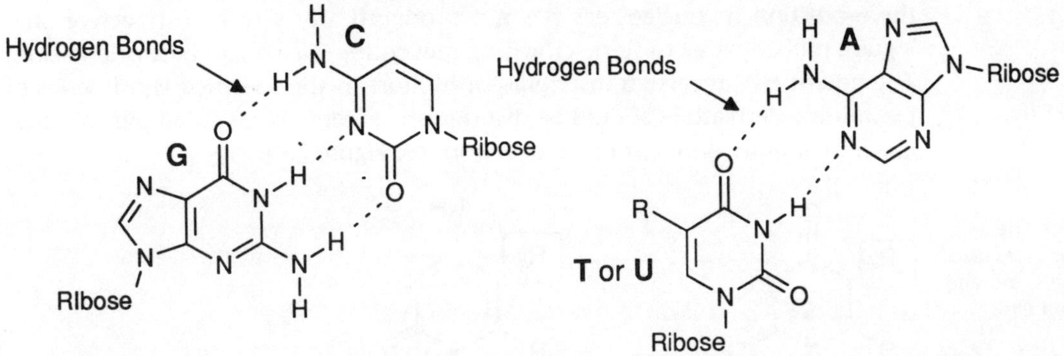

Figure 5.7 Hydrogen bonding patterns between nucleobases. These are very specific orientations contained within the complex 3-dimensional structure of base-paired nucleic acids.

Because of these neatly tailored recognition processes, there has been little scope for changing the structure of the purine or pyrimidine base in the search for nucleoside analogue therapeutics. Not only does the hydrogen-bond pattern have to be preserved, but also the scope for substituents elsewhere is limited, because subtle electronic effects within the molecules can affect the protonation or tautomeric form of the base, crucial to the recognition process. As this chapter develops, the abundance of unchanged cytosine, thymine, and guanine structures in active molecules will be evident. Adenine analogues are problematical, due to two factors: they are generally either readily metabolized by adenosine deaminase to their inosine derivatives (Figure 5.8), which are generally inactive (though ddI is a notable exception), or they interfere with adenosine metabolism which has many other important biochemical functions.

Figure 5.8
Conversion of adenosine to inosine derivatives by the action of adenosine deaminase.

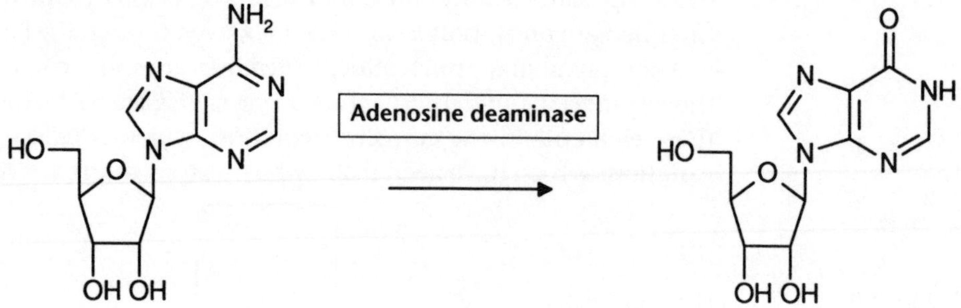

Base variations

Whilst conserving the key features required for recognition, there has been some scope for variation in the nucleobases; some substituents on carbon and minor changes in the heterocyclic backbone have produced activity. The 5-position of the pyrimidines uracil and cytosine, has been the most extensively modified, with a variety of small substituents which include halogens, alkyl, alkenyl and alkynyl groups (Figure 5.9). The use of such simple substitutions has produced active compounds both with and without altering the structure of the attached sugar. The activity of some compounds, such as BVaraU and zonavir, is explained by their high specificity for activation by viral kinases. Substitution at the 6-position in nucleosides has not produced any significantly active analogues, probably due to steric crowding preventing the correct base orientation. The potent HIV reverse transcriptase inhibitors in the so-called HEPT series of pyrimidine derivatives should be distinguished from glycosylated pyrimidines, given their quite different mode of action (see Figure 7.8).

Figure 5.9 The pyrimidine bases uracil and cytosine, showing substitution sites.

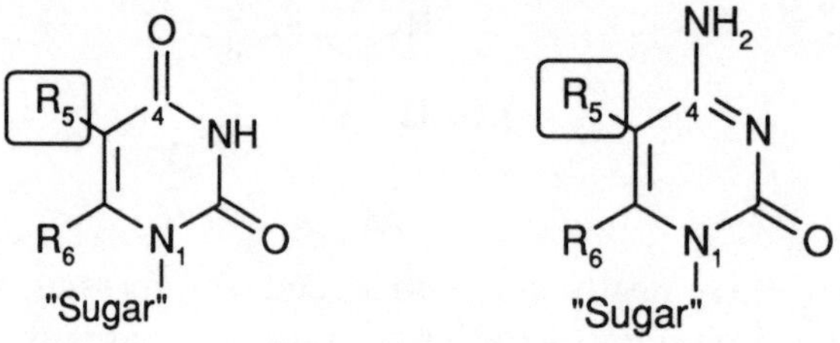

Purine bases have shown less scope for variation, although the pattern and number of nitrogen substituents has been altered (Figure 5.10), producing some active examples such as 3-deaza- or 8-aza-analogues. Chemically, due to a lack of functionalizable carbon centres, there is less scope for altering the structure without either affecting the hydrogen bonding elements or adding bulk to the 8-position which will give restricted rotation in the nucleoside molecule.

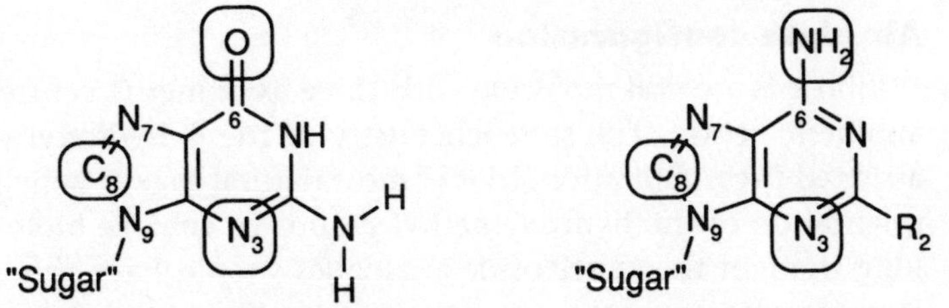

Figure 5.10 The
purine bases adenine
and guanine, showing
substitution sites.

Figure 5.10 The purine bases adenine and guanine, showing substitution sites.

There are examples of molecules which act as purine or pyrimidine mimics, such as substituted imidazoles or triazoles, notably the natural product ribavirin (Figure 5.11), which is thought to act as an adenosine mimic. These heterocycles can present suitable hydrogen bonding templates in a similar manner to either of the natural purine or pyrimidine patterns.

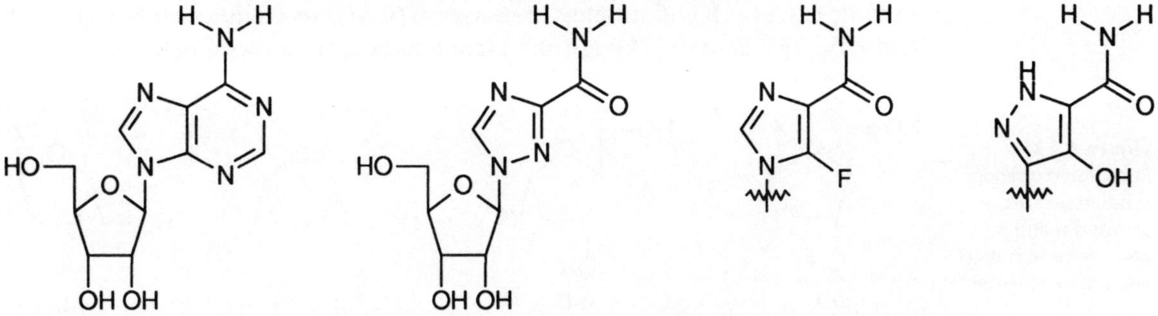

Figure 5.11 Adenosine and a triazole mimic, ribavirin, with related imidazole and pyrazole base variants.

Sugar variations

The fundamental biochemical processes of evolution and life itself are based on the highly specific orientations resulting from functionalization of the sugar D-ribose. This acts as a precisely configured scaffold which presents the hydroxymethyl group, or its triphosphate derivative, the 3'-hydroxyl and the nucleobase in highly specific orientations for the nucleic acid processing enzymes. In the nucleic acid, the sugar also acts as a specifically oriented connector between the interlinking phosphate groups, controlling the helical structure. These exacting configurations evolved with D-ribose, and its 2-deoxy version, in their 5-membered ring, or **furanose** forms (Figure 5.12).

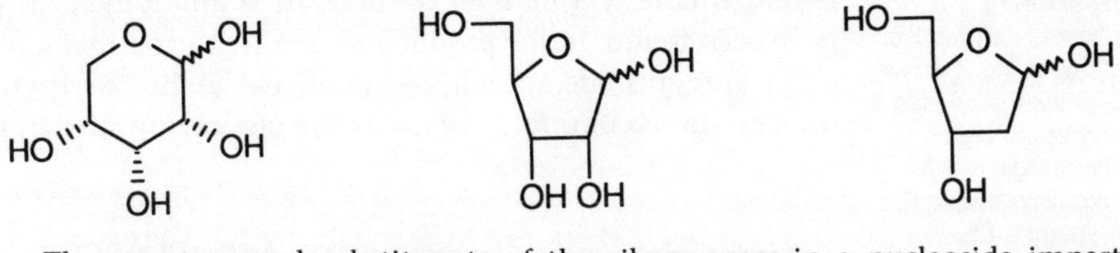

Figure 5.12 D-Ribose: pyranose and furanose forms and 2-deoxyribofuranose.

The structure and substituents of the ribose sugar in a nucleoside impart constraining features which are vital to the function of the molecule, yet still provide the basis for much variation in analogue design. The **absolute configuration**, due to **asymmetric carbon centres** in this **chiral** molecule, and the **anomeric configuration** of the base–sugar linkage are constraints which usually have to be met.

Absolute configuration

D-Ribose is a **chiral** molecule with three asymmetric centres in addition to the anomeric centre. The stereochemistry of the 4'-hydroxyl is in the historically assigned D-configuration, that of most natural sugars, which defines the specific orientation of the hydroxymethyl group. To achieve biological activity, it was long thought that nucleoside analogues should have this same configuration. However, more recent findings have led to the discovery of numerous molecules with the unnatural or L-configuration, which have potent anti-viral activities. The stereochemistry of furanose sugar molecules is shown in Figure 5.13. Generally, unnatural enantiomers of active nucleosides, such as AZT, have proven inactive, though examples where the activity lies solely in the L-enantiomer are emerging, such as 4391W which has potent anti-HIV/HBV activity. In some cases both enantiomers are active; 3TC and its enantiomer have potent anti-HIV/HBV activity, though the L-configured 3TC shows a better selectivity profile; similar correlations can be made with D- and L-ddC

Figure 5.13
Substituted furanose sugar molecules, showing α and β anomers with both D and L stereochemistry.

Anomers

The bond between the sugar and the base (or **glycosyl linkage**) has two potential orientations, termed **anomers**. The beta anomer has the base oriented on the same face as the hydroxymethyl group whereas the opposite orientation is termed alpha. Only the beta linkage is found in nature. This can cause problems in synthesis, as the processes for producing nucleoside analogues generally rely on the chemical glycosylation of an intact base, which usually produces both anomers. On an exploratory scale, the presence of around 50% of the unwanted (and virtually always inactive) anomer can be ignored, but it is a significant problem on a large scale or where pure material is required. The development of drugs has been delayed due to shortcomings in producing pure beta anomers, although latterly a number of innovative solutions to this particular problem have been found. In the production of AZT, the natural nucleoside thymidine is a key intermediate. Given the poor availability of thymidine from natural sources, this is commonly made by the process outlined in Figure 5.14.

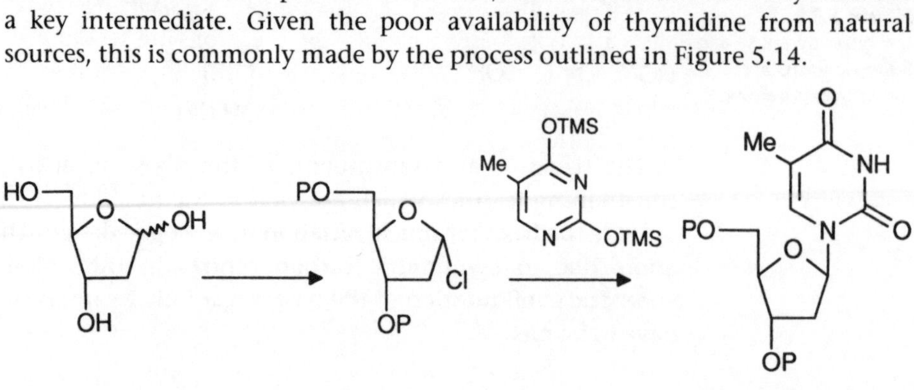

Figure 5.14
Commercial production of thymidine. 2-deoxy-D-ribose is converted to the protected chlorosugar which crystallizes as a single α-anomer. This glycosylates silylated thymine with clean inversion of configuration, to give only the protected precursor to β-D-thymidine.

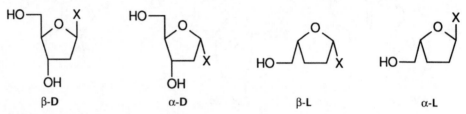

2' and 3' substituents

The other two stereocentres of the D-ribose in a nucleoside are the hydroxyls on the α-face of the sugar ring in the 2' and 3' positions. These two carbon centres have provided the most scope for significant variations. In 2-deoxyribose itself, one of these hydroxyls is absent, but the 3'-oxygen still provides the important link in the polymeric nucleic acid structure. The 2' and 3' positions combined offer four potential sites of substitution, which gives a vast potential for possible variations as summarized in Figure 5.15, examples of which will be seen throughout the book.

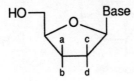

Figure 5.15 D-configured furanose sugar molecule, showing four potential sites for substitution. The most common substituents to confer activity include H, F and N_3.

2',3'-dideoxy sugars

Removal or substitution of both the 2'- and 3'-hydroxyls (Figure 5.16), produces compounds which are obligatory chain terminators of the polymerization process, a key feature for many antiviral compounds, such as the anti-HIV dideoxy nucleosides and their didehydro derivatives containing an olefinic linkage.

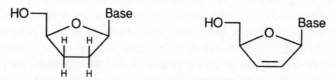

Figure 5.16 Dideoxy and dideoxy-didehydro furanose sugars.

2'-deoxy and arabino sugars

The 3'-hydroxyl is sometimes retained (Figure 5.17), usually in combination with other ring substitutions and/or base modifications. In these cases chain termination is not possible, so selectivity must be achieved in the enzyme kinetics or phosphorylation effects. In the former case, the viral polymerase must be better inhibited than the host polymerases by the triphosphate; in the latter case this is not so important, as the molecule would only be phosphorylated in infected cells. In practice, the synthetic analogues fall in between these extremes, so many 2'-deoxy or arabino nucleosides, with a few notable exceptions, tend to be poorly selective or toxic.

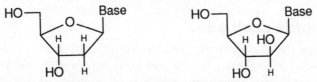

Figure 5.17 2'-Deoxyribo and arabino sugars.

Heterocyclic ring modifications

The furanose sugar is a feature of the natural nucleosides. The ring oxygen has been replaced, with consequent alterations in the biological profile, by methylene (carbocyclics), sulphur (thio) or nitrogen. There are also examples of active pentacyclic-based nucleosides where the site of the heteroatom has been moved or when two heteroatoms are in the ring. Similarly, inhibitory molecules have also arisen when the number of atoms in the ring has been expanded to six or contracted to four (Figure 5.18).

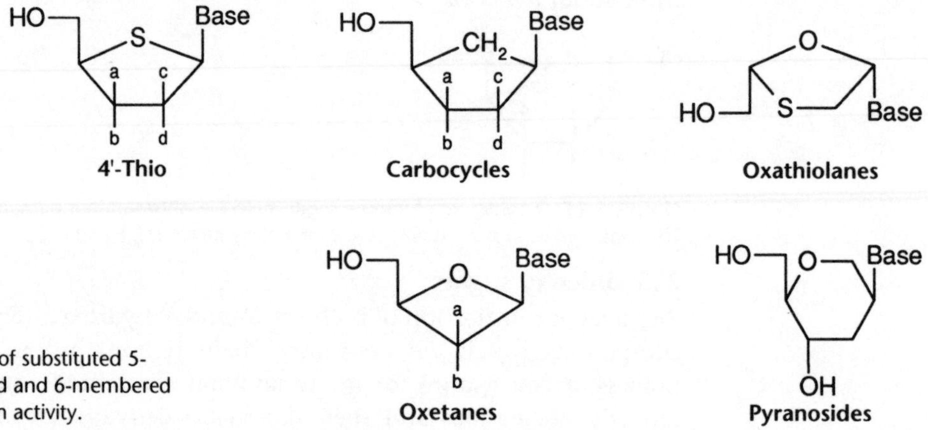

Figure 5.18 Examples of substituted 5-membered, 4-membered and 6-membered 'sugar' rings which retain activity.

Acyclic nucleoside analogues

The most radical modification is not to bother with a sugar ring at all, but to have an open chain from the base which includes a terminal hydroxymethyl group. This is best exemplified by acyclovir, in which the 'sugar' is reduced to the bare minimum of atoms (Figure 5.19). Remarkably, such a truncated structure can still retain activity, whilst making the molecule a mandatory chain terminator. Variations on this theme include analogues with branched chains such as ganciclovir and penciclovir, which, like most other molecules in the class, rely chiefly on virally encoded functional proteins for their phosphorylation. Many active phosphonate nucleosides also have straight chain or branched chain acyclic structures (Figure 5.24).

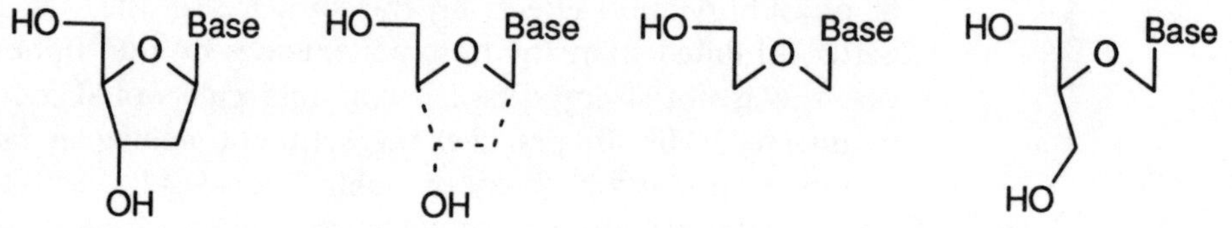

Figure 5.19 Parent nucleoside, removable substructure and acyclic variations.

Prodrugs of nucleoside analogues

One of the problems with nucleoside analogues can be their poor absorption after oral administration. This can been improved by the formation of prodrugs of the nucleoside analogue, which itself might be regarded as a prodrug of the active triphosphate form. Better absorption may only be one factor in the over-all improvements in pharmacokinetics sometimes achieved by prodrugs. After absorption, endogenous enzymes are required to modify the prodrug in order to produce the nucleoside form of the molecule, either by cleavage of a substituent or biochemical modification of the structure, a process which could take place intracellularly. There are various ways of achieving this, examples of two of the more important ones are as follows.

Valaciclovir is the valinyl ester of acyclovir, which is better absorbed than the parent drug into the GI tract, where it is cleaved by host enzymes to yield the natural amino acid valine plus acyclovir (Figure 5.20). Because the ester is better absorbed, smaller and less frequent doses are required to maintain a therapeutic level of the drug.

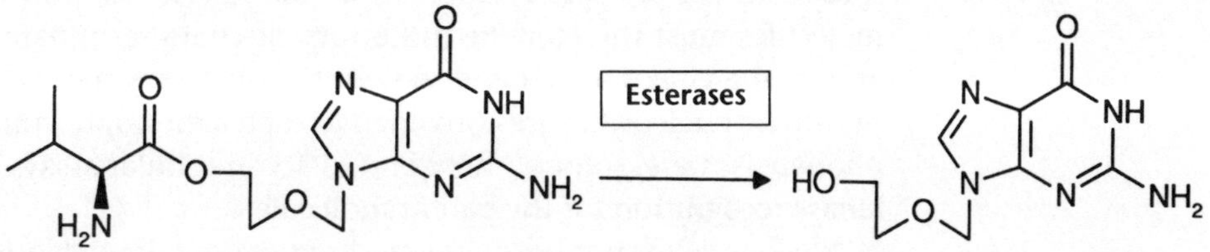

Figure 5.20
Conversion of the prodrug, valaciclovir, to acyclovir by esterases.

A different acyclovir prodrug strategy used the desoxy compound BW515U, which, like valaciclovir is better absorbed. This is converted to acyclovir by the action of the endogenous enzyme xanthine oxidase (Figure 5.21).

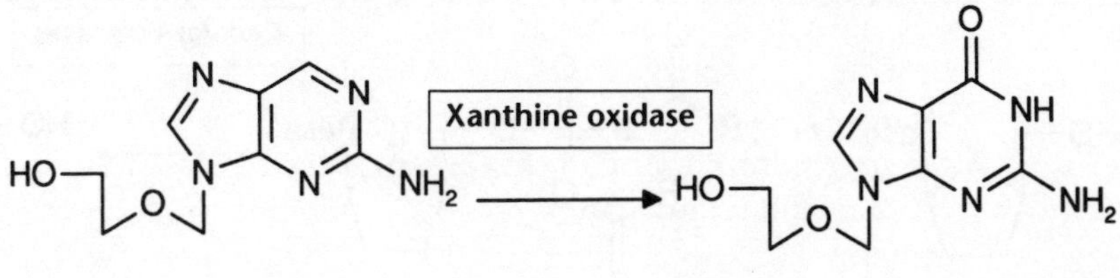

Figure 5.21 *In vivo* activation of BW515U to acyclovir by the action of xanthine oxidase.

Famciclovir requires the action of both an esterase and xanthine oxidase before it is converted to penciclovir, the proximate nucleoside analogue in the sequence, which ultimately forms the inhibitory triphosphate (Figure 5.22).

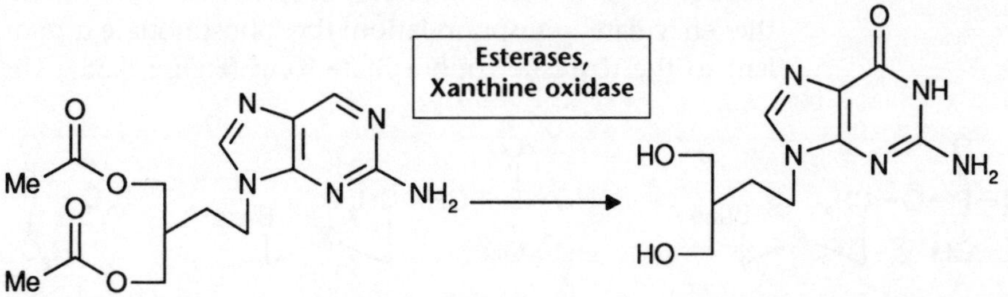

Figure 5.22 *In vivo* conversion of famciclovir to penciclovir, the nucleoside precursor to the active triphosphate form.

Nucleoside phosphates: isosteres and prodrugs

In preceding sections the necessity of nucleoside phosphorylation has been highlighted, in addition to the problems of nucleoside phosphate delivery to the intracellular site of viral replication. There have been numerous approaches to circumvent these difficulties which aim to deliver nucleoside phosphates or their isosteres into cells. Nucleoside monophosphates and their isosteres, sometimes referred to as **nucleotide analogues**, have received most attention. These are good targets for delivery, since the greatest substrate specificities are observed for this conversion and monophosphates are generally more readily processed to triphosphates by nucleotide kinases of either or both cellular and viral origin. There is a drawback in this mechanism of activation however, since the selectivity of such molecules is often reduced, given that the possibility of specific initial phosphorylation by a viral kinase has now been removed. Such molecules must therefore be inhibitors, or chain terminators, acting selectively against the viral polymerases and not the host enzymes. When some inactive or non-toxic nucleosides are converted to such a monophosphate prodrug or phosphonate isostere, some activity or toxicity in cellular assays would indicate poor kinase recognition for the parent molecule.

Nucleoside monophosphate prodrugs consist of nucleoside monophosphates (themselves phosphate monoesters) with substituents on the either or both of the available oxygens on the tribasic phosphate (Figure 5.23). These are cleaved by intracellular enzymes to release the monophosphate. To date, activation has been achieved by this method, though no serious drug candidates have emerged.

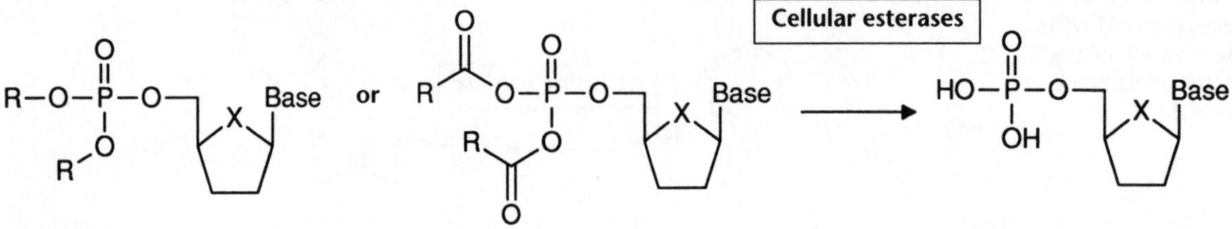

Figure 5.23 Nucleoside monophosphate prodrugs.

Phosphonates have provided useful monophosphate mimics: numerous examples have been studied extensively, many showing broad spectrum antiviral actions, with useful therapeutic windows. The phosphonate, which has a phosphorus directly bound to carbon (Figure 5.24), not in an ester linkage through oxygen, acts as a monophosphate isostere, which is a substrate for further enzymatic phosphorylation; the 'phosphonate diphosphate' being equivalent to the requisite triphosphate form (Figure 5.25). Their profiles have also

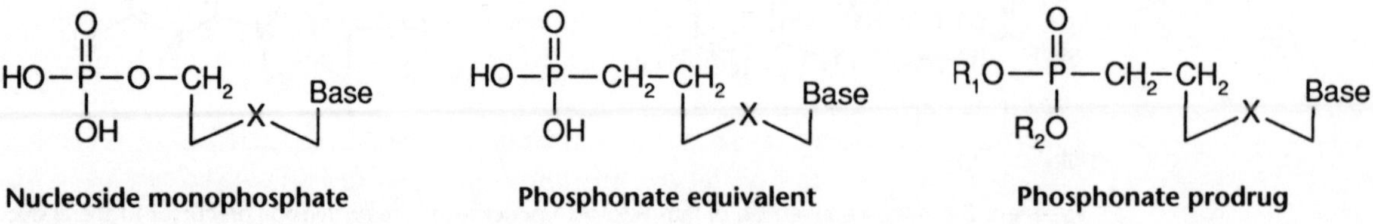

Nucleoside monophosphate   Phosphonate equivalent   Phosphonate prodrug

Figure 5.24 Nucleotide analogues.

been improved by the formation of prodrugs, the equivalent of the monophosphate esters, which liberate the phosphonate inside cells. Such molecules are of particular potential against viruses which do not encode a kinase or have mutations which have altered or deleted this enzyme, although significant activity is sometimes found against kinase encoding viruses too. A general drawback, due to the phosphorylation mechanism, is generally poorer selectivity, although the relative inhibition of viral and host polymerases (selectivity index) is wide enough in some cases to be of therapeutic benefit.

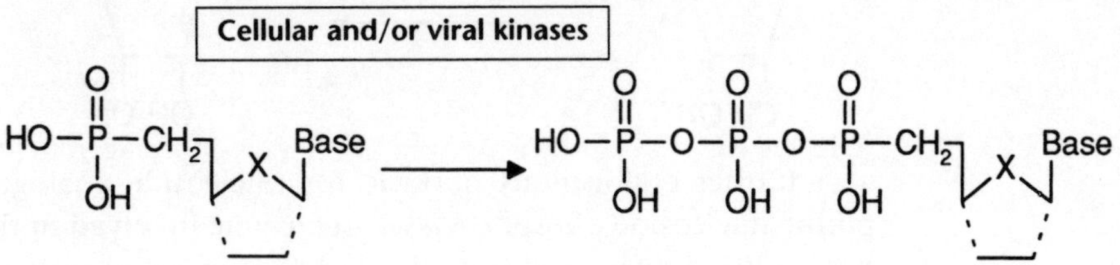

Figure 5.25
Nucleoside phosphonate isosteres and their conversion to triphosphate equivalents.

Phosphate analogues

The phosphate binding site of the nucleic acid polymerases has itself provided a binding site for therapeutic intervention; blocking this with pyrophosphate isosteres prevents nucleic acid synthesis. Intervention at this site of viral enzymes can produce selective inhibition over cellular equivalents. Such compounds have the advantage of not requiring any activation and they often show a broad spectrum of activity, though these are often compromised by poor selectivity. Foscarnet (Figure 5.26), the trisodium salt of phosphonoformic acid, despite a relatively poor selectivity, is used clinically against CMV infection, notably in cases where viral resistance emerges. This is of most use where kinase mutations occur, since polymerase mutant phenotypes may show cross resistance between foscarnet and nucleoside agents. Other pyrophophates, such as phosphonoacetic acid (Figure 5.26), act similarly, though these lack even the selectivity of foscarnet.

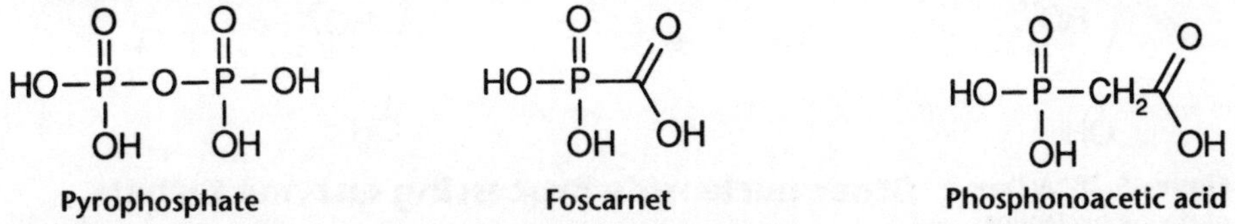

Figure 5.26
Pyrophosphate and two of its mimics: foscarnet (phosphonoformic acid) and phosphonoacetic acid.

Nucleoside metabolism and transformation

Like any xenobiotic substances, nucleoside analogues are potentially subject to many kinds of modifications once they have been absorbed into the body, these are all part of the body's natural defences against foreign molecules. Typically glucuronide conjugates would be formed with the free 5'-hydroxyl of a nucleoside, a glycosidation process with glucuronic acid which renders the molecule inactive. In addition to these mechanisms, endogenous nucleoside processing enzymes can also modify the administered compound. The consequences of the action of one such enzyme, adenosine deaminase, was mentioned in Figure 5.8; a similar deamination process can also modify the actions of active cytidine analogues, when cytidine deaminase converts the molecule to its often inactive uridine equivalent (Figure 5.27)

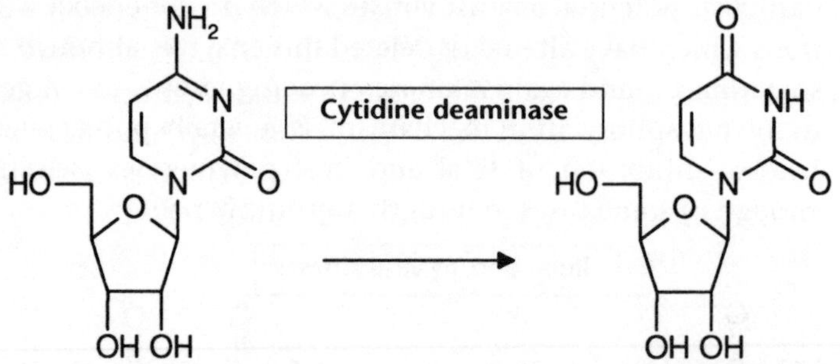

Figure 5.27 Action of cytidine deaminase.

A further endogenous obstacle for nucleoside analogues is the action of purine nucleoside phosphorylase, an enzyme involved in the natural process of nucleoside cleavage and transglycosylation of sugars onto other bases. This can lead to complications if the cleaved nucleobase is toxic or has unfavourable interactions: for example, BVaraU is cleaved *in vivo* (Figure 5.28) to give bromovinyl uracil which potentiates the cytotoxity of 5-fluoro uracil (an anti-cancer drug) in a potentially lethal cocktail. The use of carbocyclic and to a lesser extent thio nucleosides may resolve this problem; in the former case the enzyme cannot function due to the inherent structure and in the latter, a strengthened glycosidic linkage is less prone to cleavage. Interestingly unnatural (or L-) nucleosides are poor or unrecognizable substrates for both PNP and the deaminase enzymes, thus providing a further solution to the problem.

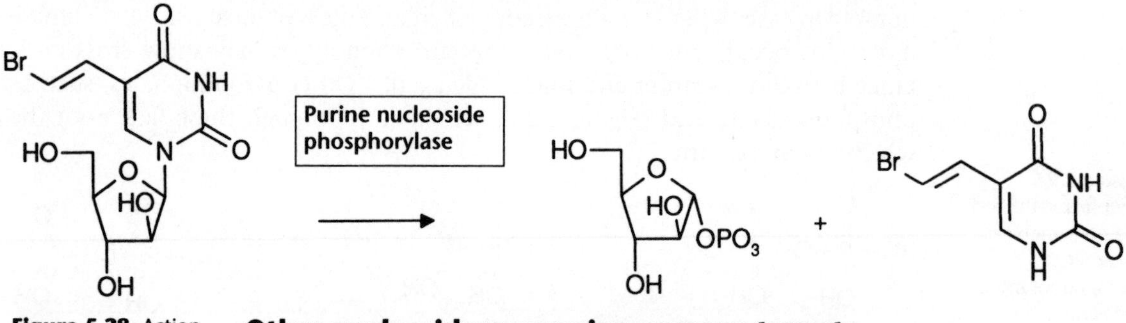

Figure 5.28 Action of purine nucleoside phosphorylase on BVaraU.

Other nucleoside processing enzyme targets

Antiviral activity is not only manifested in compounds which chain terminate or inhibit the function of polymerase enzymes. Some nucleoside analogues, which show antiviral activity, such as ribavirin, act by inhibiting other enzymes involved in the complex methods of nucleic acid processing and biosynthesis. However, a drawback of such targets is that the enzymes are generally of host, not viral origin, so the selectivity of such inhibitors tends to be somewhat lower than those which target viral polymerases. Herpes virus ribonucleoside reductase has received some attention as another viral nucleic acid processing target.

Viral enzymes may deal with nucleic acid processing in forms other than nucleosides. HIV integrase, for example, catalyses the integration of viral DNA into host DNA, a process which has no mammalian equivalent.

Oligonucleotides

Nucleic acid oligomers and their modified derivatives represent a very new area of investigation with applications in antiviral chemotherapy. Such oligonucleotides consist of short single strand nucleic acid sequences which are polymers of the nucleoside monomers, often with modifications to the phosphate linkage to introduce stability. The theory behind such approaches does not rely on the disruption of an actively replicating virus, so could potentially restrict viruses to their latent phases. Oligonucleotide therapies can be broadly classified under four separate approaches: antisense, antigene, decoy and ribozymes.

Antisense nucleotides

The concept here is to produce an RNA sequence which is complementary to single-stranded viral RNA, capable of producing double stranded (ds) RNA. By forming the dsRNA duplex the viral genome is effectively neutralized, thus shutting down the viral RNA replication. The dsRNA may induce other host responses; it might either be digested by host nucleases or trigger a localized interferon response. Antisense nucleotides are either produced by a synthetic gene introduced to a cell or, more commonly, chemically synthesized, often with modifications in the phosphate linkers. Phosphorothioates, in which the phosphates are substituted with a sulphur, are under early stages of clinical evaluation.

Antigene methods

The antigene approach, which is somewhat less developed than the related antisense strategy, targets DNA rather then RNA. The aim is to form a triple helix of DNA where by some stabilized or modified nucleotide is able to bind to specific regions in the DNA double helix, forming the stable hybrid form, thus preventing transcription.

Decoy methods

Specific regions of viral nucleic acids are the targets for virally encoded regulatory proteins which affect transcription, such as the HIV TAT–TAR interaction. By using a synthetic decoy corresponding to the specific nucleic acid sequence which binds the regulator, the interaction can be disrupted, potentially inducing a strong antiviral effect.

Ribozymes

Ribozymes are RNA molecules that are able to induce self cleavage or cleavage of other RNA sequences at specific sites. Whilst practical applications as antiviral agents are perhaps somewhat distant, the possibility exists for producing synthetic ribozymes (themselves oligo(ribo)nucleotides) which target and cleave specific viral nucleic acid sequences. Therapies for both RNA and DNA viruses could be possible; in the former case, genomic RNAs would be specific cleavage targets, whilst mRNA transcripts from DNA viruses could also be selected and cleaved.

6 | Herpes viruses

Infections by herpes viruses are amongst the most common and easily transmitted viral conditions. Numerous distinct viruses have been identified (Table 6.1) and the treatment by chemotherapeutic agents of diseases caused by some of these, such as HSV-1, HSV-2 and VZV, has produced substantial clinical benefit. Whilst the earliest true antiviral treatments were poorly selective topical treatments for herpes conditions, the entry of acyclovir as the first truly selective and efficacious antiviral drug was a milestone in antiviral chemotherapy.

Table 6.1 Significant herpes viruses affecting man.

Subfamily	Virus	Abbreviation
Alphaherpesvirinae	Herpes simplex virus type 1	HSV-1
	Herpes simplex virus type 2	HSV-2
	Varicella zoster virus	VZV
Betaherpesvirinae	Cytomegalovirus	CMV
	Human herpesvirus type 6	HHV-6
	Human herpesvirus type 7	HHV-7
Gammaherpesvirinae	Epstein-Barr virus	EBV

Herpes viruses are relatively complex, some 120 to 200 nm in diameter with an enveloped structure and an icosahedral nucleocapsid. The genomic linear double-stranded DNA has a molecular weight between 85,000 and 160,000 kDa and codes for about 100 polypeptides. Whilst the functions of most of the encoded proteins are poorly understood, a number of them are well characterized and include enzymes which provide excellent targets, or have vital implications, for chemotherapy. These include a **DNA polymerase** essential for replication, a **protease** enzyme which has strong homology across the family and, significantly for HSV-1, HSV-2 and VZV, a **thymidine kinase**. Anti-herpes virus agents often show a broad spectrum of activity across the family, a consequence of structural similarities between the functional proteins, although the presence or absence of the thymidine kinase is usually a key factor in selectivity.

Herpes simplex viruses

Infections due to the closely related herpes simplex viruses types 1 (HSV-1) and 2 (HSV-2) manifest themselves in similar ways. They all could be caused be either virus, though HSV-1 primarily affects the upper part of the body whereas HSV-2 is more commonly associated with genital infections. The most commonly observed HSV-associated condition is cold sores. Most people are first exposed to HSV in childhood, often as the result of a kiss from an older person. The resulting infection is often asymptomatic, though herpetic gingivostomatitis may result. The infection will heal, but the virus will then remain dormant until some factor induces reactivation, the unpleasant consequence of which will be cold sores.

Skin contact provides another means of HSV transmission; HSV in oral secretions often infects healthcare workers and dentists. Herpetic lesions themselves can be highly infective; close contact sports such as rugby, wrestling and the martial arts provide an effective vehicle for the dermal syndrome known as herpes gladiatorum. Like the agents for many other infectious diseases, herpes viruses are rapidly spread by sexual activity; mobile body fluids and intimate dermal association promote rapid transfer of highly infectious virions. There are suitable animal models of HSV infection, whereby the pathogenic effects of the virus can be seen, for example, in guinea-pigs.

Pathogenesis, latency

HSV infections are primarily channelled through the skin and mucous membranes; here the virus replicates, causing cellular lysis and the formation of vesicles. Outward signs of infection are shallow ulcers in the mucous membrane due to rupture of the vesicles, these remain intact for several days in dermal infections, before they crust over and heal. The vesicle contains intranuclear inclusion bodies, the site of virus replication and a clear fluid containing large numbers of virions. In rare cases, the virus can further spread to the central nervous system causing meningitis or encephalitis, but another, more subtle mechanism, plays a role in the long-term consequences of HSV infection. Virions travel from the site of primary infection to root ganglia via the connecting sensory nerves. Here the virus remains latent until reactivated, whence it travels in the reverse direction to induce a recurrent lesion (Figure 6.1). The mechanisms and reasons for such reactivation are poorly understood, but a variety of stressful factors may be implicated.

Treatment of HSV

In discussing the treatment of HSV infections care must be taken to distinguish between **therapy**, by a true antiviral agent or **alleviation** of the symptoms, on which some clinically used preparations rely. Furthermore, even after successful therapy or prophylaxis of a lesion, the root ganglia are not cleared of infection, so a distinction between the **treatment** rather than a **cure** must be also be

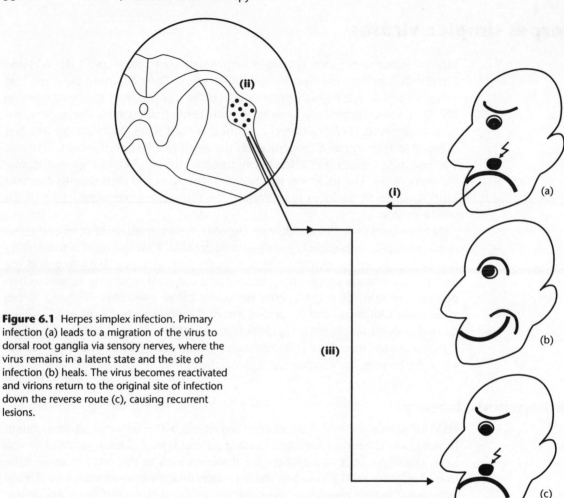

Figure 6.1 Herpes simplex infection. Primary infection (a) leads to a migration of the virus to dorsal root ganglia via sensory nerves, where the virus remains in a latent state and the site of infection (b) heals. The virus becomes reactivated and virions return to the original site of infection down the reverse route (c), causing recurrent lesions.

emphasized. Ideally, reactivated HSV should be challenged with chemotherapeutic agents before vesicles are formed; hence the necessity for the application of drugs at the associated 'tingle' phase of cold sore symptoms.

True antiviral therapy has been successfully achieved by a number of agents, most of which inhibit the replication of viral DNA. The agents of clinical significance are nucleoside analogues, although other molecules have also exhibited distinct *in vitro* activity.

Nucleoside analogues

Acyclovir

Acyclovir, as a treatment for HSV infections, was the first example of a genuinely selective antiviral agent. It has profound effects on the viral DNA polymerase function through obligatory chain termination and competitive inhibition, acting as its triphosphate form. Acyclovir is a substrate for viral thymidine kinase, but not for any host kinases, which accounts for the overall selectivity and is a

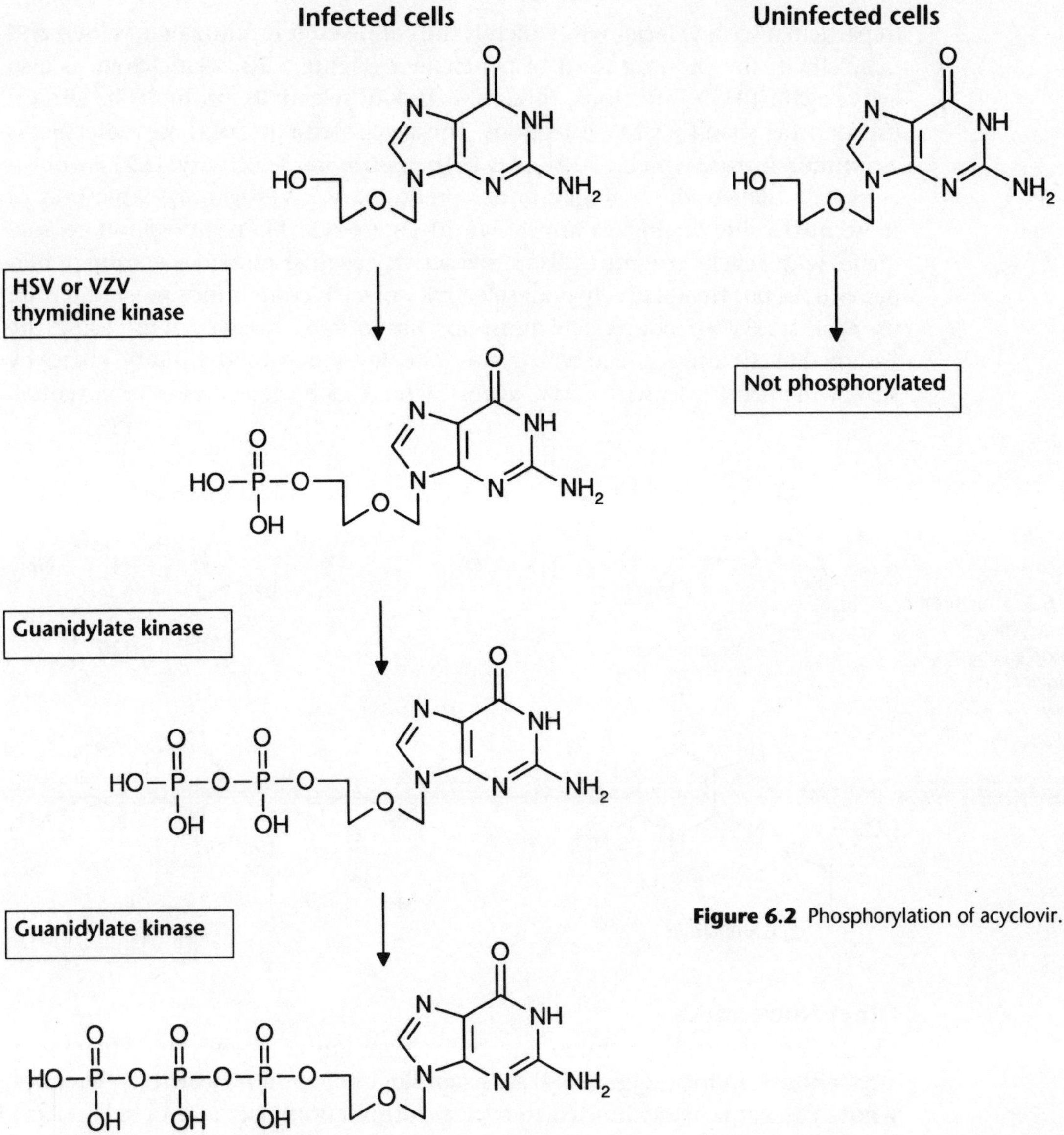

Infected cells **Uninfected cells**

HSV or VZV
thymidine kinase

Not phosphorylated

Guanidylate kinase

Guanidylate kinase

Figure 6.2 Phosphorylation of acyclovir.

reason for the excellent safety profile, as the drug can only be phosphorylated to its active form in infected cells (Figure 6.2). Paradoxically, micro-injection of acyclovir monophosphate into an uninfected cell can result in cellular toxicity. This occurs because enzymes catalysing the second and third additions of phosphate molecules are less selective than the nucleoside kinase, thus significant levels of ACV-triphosphate are produced, which then interferes with the action of cellular polymerases.

Prodrugs and Other Acyclic Nucleoside Analogues

One disadvantage of acyclovir is its relatively poor absorption and pharmacokinetics. This is a minor drawback that has been overcome by the use of prodrugs, in particular with valaciclovir, which is the valine ester prodrug of acyclovir and famciclovir, the prodrug form of penciclovir (Figure 5.20). Ganciclovir is also active against HSV infections, though its lack of selectivity prohibits its clinical use for other than for CMV infections. The acyclic structure of these molecules is a common feature which contributes to their antiviral selectivity: H2G is another acyclic nucleoside analogue under preliminary investigation. Structures of these nucleoside analogues are shown in Figure 6.3. Many phosphonate analogues with acyclic structures also show activity against a broad spectrum of herpes viruses, but the relatively poor selectivity of such compounds precludes their use against HSV infections. The phosphorylation step, which such molecules are designed to circumvent, can be selectively achieved by viral thymidine kinase by HSV, which contrasts with CMV, against which such molecules have potential.

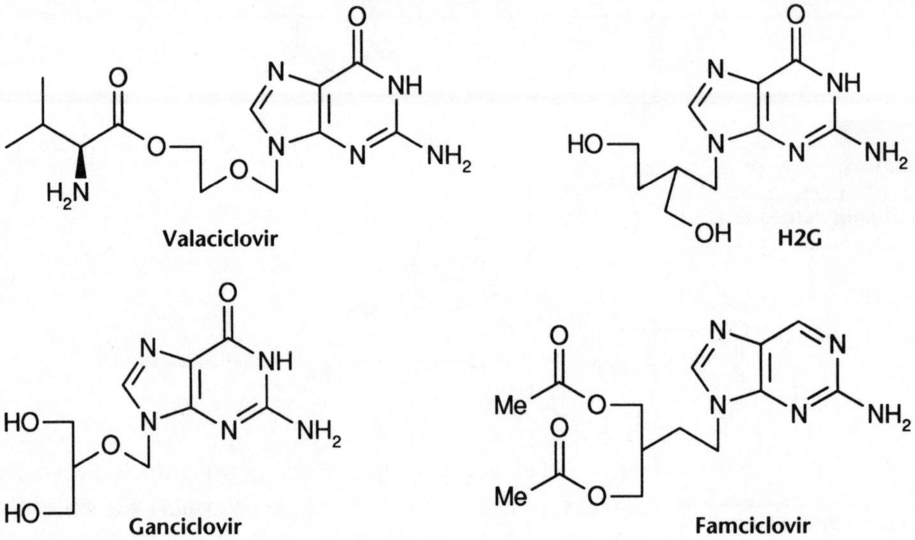

Figure 6.3 Structures of some acyclic nucleoside analogues, active against HSV infection.

Other Nucleosides

Idoxuridine and trifluorothymidine, both thymidine analogues, and vidarabine, an adenosine mimic, (Figure 6.4) are examples of older, non-selective, therapies, whose clinical usage is limited to topical formulations, because of side effects. These compounds are phosphorylated by the viral kinase, ultimately producing inhibitors of the viral DNA polymerase. However, they are also phosphorylated in uninfected cells by host kinases and are substrates for human DNA polymerases, which limits their selectivity.

A number of more recent nucleoside analogues with intact sugar rings, such as lobucavir, GGR61192, and 304C, have entered development as potential anti-HSV agents. Their activities, like the acyclics, are due to selective phosphorylation and/or selective polymerase inhibition, though they have not been obligate chain terminators. Despite the apparent promise of some such agents in animal models, many of which have shown superior efficacy and pharmaco-kinetics to acyclovir, most have been withdrawn because their safety profiles have not matched the standard of that drug.

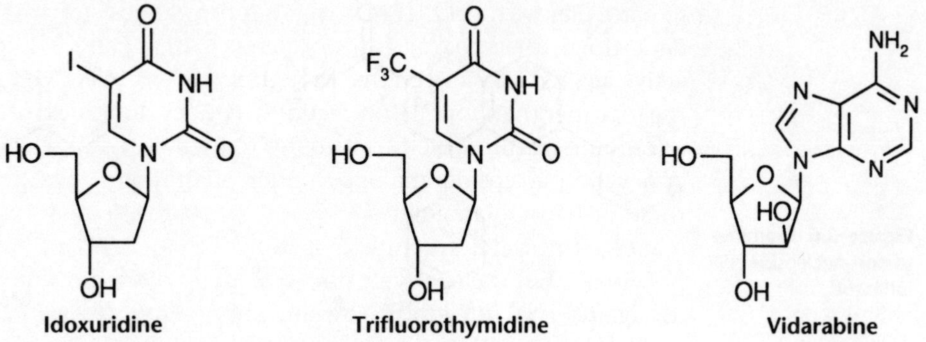

Figure 6.4 Nucleosides used in topical formulations.

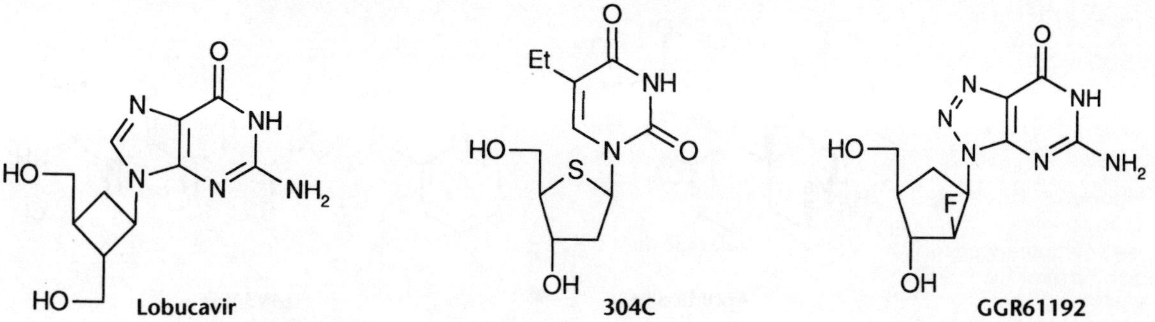

Figure 6.5 Recent nucleoside analogues with intact sugar rings.

Other Agents

The pyrophosphate analogue foscarnet (Figure 5.25) is active against some HSV infections. Whilst its toxicity profile prevents widespread use, it has applications when clinical resistance to nucleoside analogues is prevalent. Many non-nucleoside inhibitors have been discovered through screening, but few have warranted serious interest to date. Some of these, such as 3-quinolinecarboxamide and mappicine ketone, have been the subject of extensive chemical manipulation of the basic structures, often producing molecules with potent inhibitory effects, but which have fallen down due to poor selectivity or other adverse effects. Often the mode of action of such compounds is unknown.

The natural product aphidicolin (Figure 6.6) has prompted research into investigating some structural analogues as potential general herpes virus therapeutics. Aphidicolin inhibits both cellular DNA polymerase-α and the viral polymerase, although the former action makes the compound cytotoxic. However, given that no kinase activation of the molecule is required, it provided an attractive basis for chemical manipulation, with a goal of producing structural analogues which might exhibit selective inhibition of viral rather than host polymerases. Interestingly, some viral polymerase mutants, resistant to acyclovir and foscarnet, were found to be hypersensitive to aphidicolin. In spite of this, efforts waned when little selectivity could be achieved by chemical manipulation.

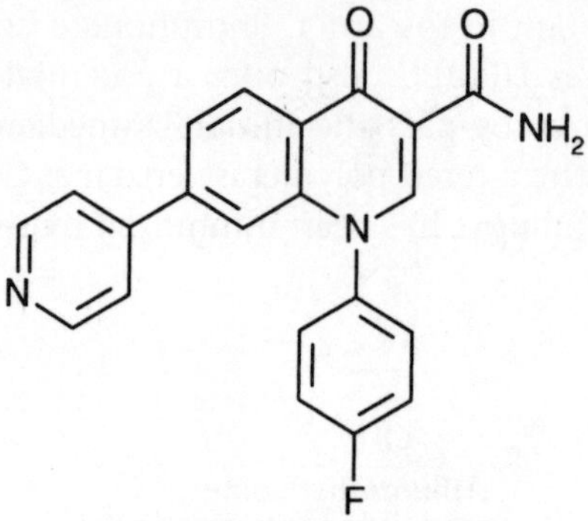

Figure 6.6 Examples of non-nucleoside HSV inhibitors.

3-Quinolinecarboxamide

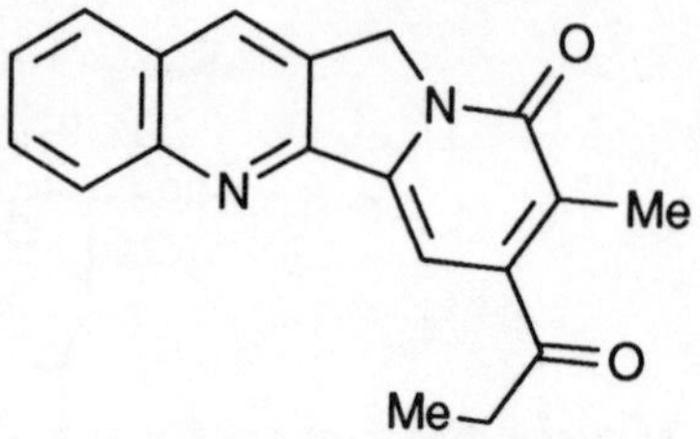

Mappicine ketone

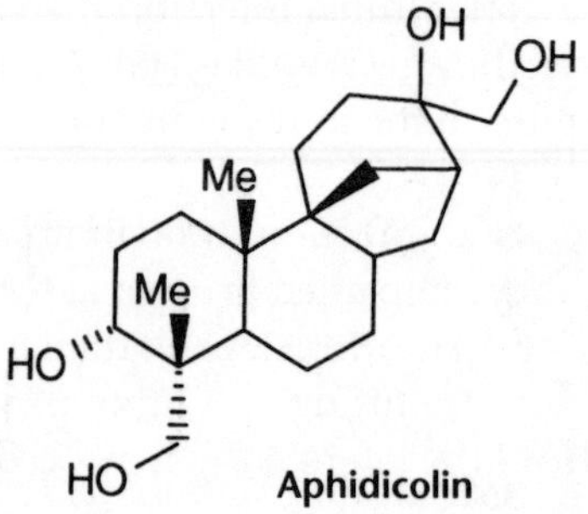

Aphidicolin

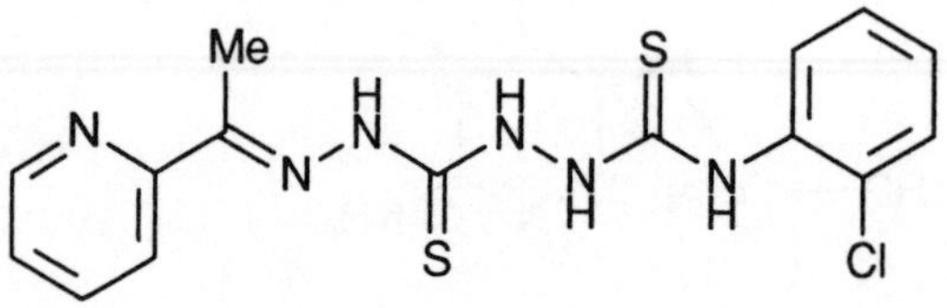

BW348U

The ribonucleoside reductase inhibitor BW348U (Figure 6.6) provided an interesting combination with acyclovir. The HSV-encoded ribonucleoside reductase enzyme catalyses the reduction of ribonucleosides to 2'-deoxyribonucleosides, essential for the formation of new copies of viral DNA. 2'-deoxyguanosine is the natural substrate which acyclovir mimics, production of which is inhibited by BW348U. Thus by depleting 2'-deoxyguanosine, which is a competitive substrate with acyclovir and its phosphorylated metabolites, the activity of the drug is potentiated.

HSV resistance to chemotherapy

With herpes simplex viruses, the evolution of drug resistant phenotypes does not appear to be clinically very significant. Resistant strains do not seem to maintain levels of pathogenicity, although the issue might be more important in immunocompromised patients.

Clinical isolates and, especially, laboratory mutant strains can be resistant to acyclovir and other nucleosides. Many of these strains have deleted or altered the thymidine kinase, which makes them unable to phosphorylate nucleosides. TK function is not vital to the survival of HSV, though it may be able to increase replication rates. *In vitro* methods have produced acyclovir resistant phenotypes, which have an alteration in the function of the DNA polymerase. These are potentially a bigger problem, but so far appear to be of little clinical significance. When acyclovir resistance causes significant problems, foscarnet is used; however, mutants where acylovir resistance is due to altered DNA polymerase function can also be resistant to foscarnet. Phosphonate isosteres of nucleoside mono-phosphates, such as HPMPC, also offer a potential solution to the TK altered mutants, since they by-pass the initial TK-mediated phosphorylation step, but not those with the altered polymerase enzyme. Complications due to the poor selectivity of inhibitors has been minimized by the use of topical formulations in some cases.

Varicella zoster virus

Varicella zoster virus (VZV), responsible for **varicella** (or chickenpox) and **herpes zoster** (shingles, or hives), is closely related to the other *alphaherpes virinae*, HSV-1 and HSV-2. Like these viruses, after initial infection the virus will lie dormant to perhaps be reactivated later; however unlike HSV-1 and 2 viruses, VZV is acquired via the respiratory route, prior to dissemination in the blood stream, leading to the characteristic symptoms.

Varicella is a common childhood infection, though it can also affect adults, manifested by a feverish illness accompanied by dermal features—macules that rapidly become papules which turn into vesicles and then to crusts that are shed from the skin. The infection is self-limiting and immunity is conferred to the virus. However, reactivation from the latent stage may occur in later life as herpes zoster.

Herpes zoster (from the Latin for belt or girdle) typically occurs as a thoracic rash in later life. This can be a debilitating condition and usually only occurs once, though post-herpetic pain is a serious problem. Even less is known about the mechanism of this reactivation than with HSV-1 and 2, although its effects are most commonly seen in immunocompromised patients and in the elderly, whose immunity is naturally in decline. Due to the specific cause of the attack, zoster is not transmitted from a varicella patient, although a primary varicella infection can be caught from a person with zoster.

Treatments for VZV infection

Given the close similarities of VZV and HSV-1 and -2, the majority of the agents discussed for the latter two are also active against VZV. Acyclovir has been the drug of choice for VZV conditions, although relatively large and frequent doses are required to maintain a therapeutic concentration, reflecting the lower potency of the drug against VZV compared with HSV. VZV thymidine kinase does not phosphorylate acyclovir as well as the HSV enzyme. The launches of famciclovir and valaciclovir offer a lower and less frequent dosing regime due to superior pharmacokinetics. Other, more potent, agents have been discovered for VZV infection, including the 5-substituted uridine analogues 882C, BVaraU, cBVdU and BMS 181,185. The propynyl or bromovinyl substituents make the molecules particularly good substrates for viral kinases, especially that encoded by VZV; indeed 882C is a highly selective agent for VZV (it is inactive against HSV-1 and HSV-2). This potent selectivity between the herpes viruses is due to 882C-monophosphate being an excellent substrate for the second phosphory-

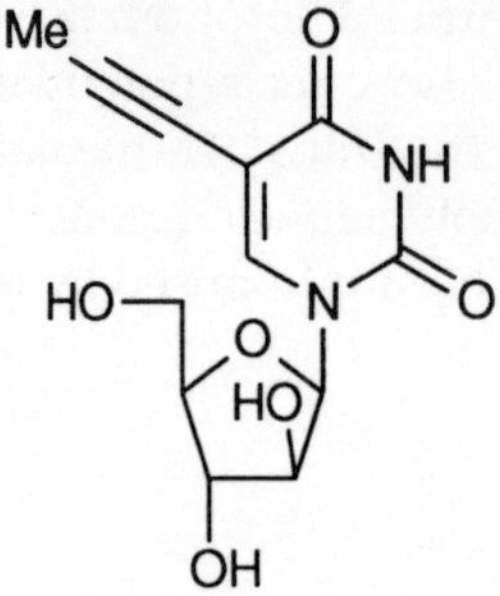

Figure 6.7 Drug candidates for the treatment of varicella zoster virus infection.

882C

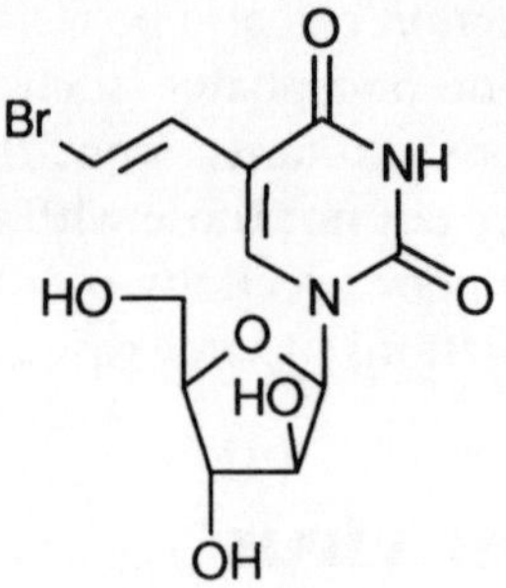

BVaraU

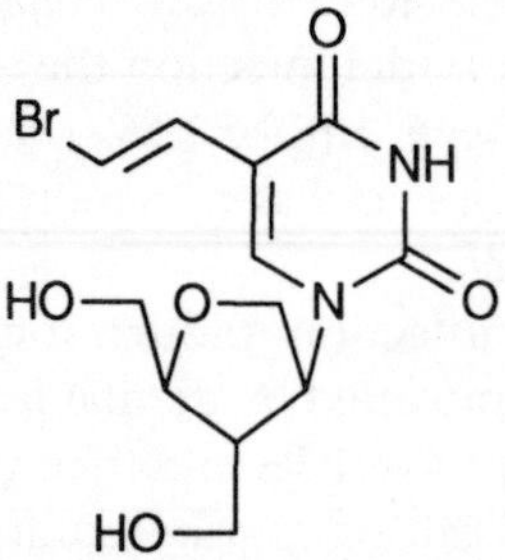

BMS181,185

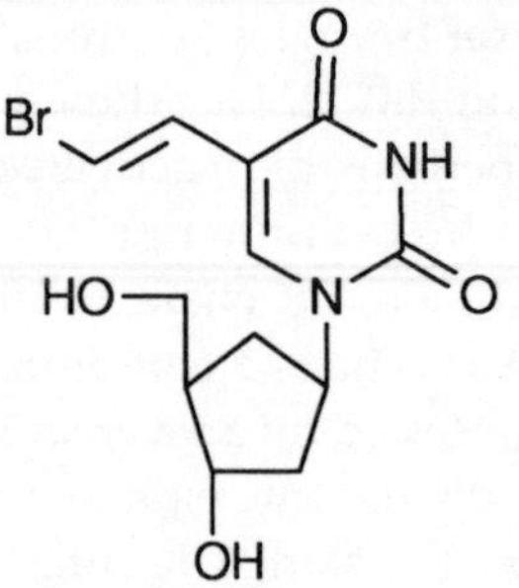

cBVDU

lation by the VZV-encoded kinase enzyme acting as the thymidylate kinase. The monophosphate is produced by the equivalent HSV-encoded kinase, but is not converted to the diphosphate; no monophosphate is produced in uninfected cells, Figure 6.8. The use of BVaraU was compromised by safety concerns over the bromovinyluracil base, a cleavage product due to the action of purine nucleoside phosphorylase (Figure 5.28). This cleavage is not possible by the non-glycosidically linked BMS181,185 and cBVDU.

Unlike for HSV infections, there is no particularly convenient animal model of VZV, other than the chimpanzee, so in developing drug candidates there is little chance to prove efficacy other than in man. In most cases, the dual activities can allow HSV to be used as a surrogate to demonstrate *in vivo* efficacy.

Vaccination against VZV infection

A vaccine against VZV infection is in the late stages of development. Even so, the clinical need for therapies will remain for some time yet, given the time scale of vaccination programmes and the widespread existence of dormant infections in the population. The symptoms of herpes zoster, the most important targets for chemotherapy, are not prevented by a vaccine, due to the origins of the disease from latent VZV infections.

Cytomegalovirus

Human cytomegalovirus (HCMV, CMV) takes its name from the 'large cells' produced when its pathogenic effects are manifested, typically producing swollen

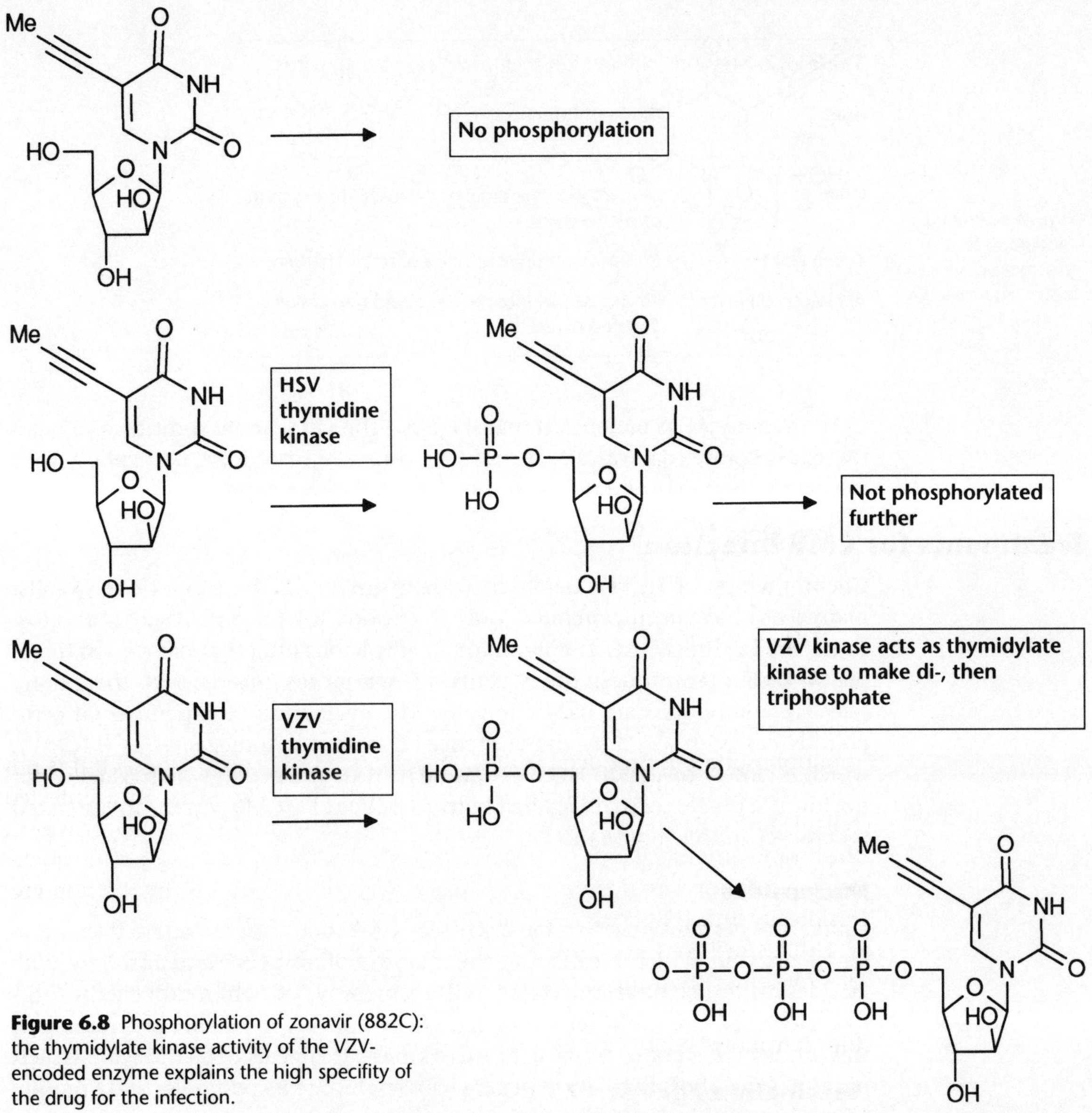

Figure 6.8 Phosphorylation of zonavir (882C): the thymidylate kinase activity of the VZV-encoded enzyme explains the high specifity of the drug for the infection.

cells with large intranuclear inclusions. Most people are infected with CMV at some stage during their lives, as summarized in Table 6.2.

In the majority of cases, CMV infections are asymptomatic throughout a person's life, although there are some important exceptions. When a foetus is infected, the virus may occasionally be the cause of neonatal disease; whilst in adolescence, it can cause symptoms akin to infectious mononucleosis (caused by Epstein–Barr virus, another herpes virus). However, the major clinically significant occurrences of CMV infection are in immunocompromised patients, when CMV-derived infections become major complications. Not only is this a problem for those with HIV infection, but also for those with deliberately suppressed immune systems associated with transplants. The ubiquitous distribution of

Table 6.2 Methods of infection by human cytomegalovirus

Age	Mode of infection
Fetus	Across the placenta
Infant	Contact with maternal body fluids (during birth, breast-feeding)
Young child	Contact with urine or saliva of other children
Adolescence onwards	Kissing, sexual intercourse, blood transfusions, donated tissues

CMV in many AIDS patients at one stage put the virus under scrutiny as a putative cause before HIV was discovered. The course of infections, particularly with respect to dormancy and reactivation is not well understood.

Treatments for CMV infection

Chemotherapy of HCMV has been widely studied in the early 1990s. Whilst many leads have been generated, only the poorly tolerated ganciclovir and foscarnet are in clinical use. The quest for an efficacious drug that has a good therapeutic profile continues; the severity of symptoms means that the poorly selective compounds are used clinically. The evaluation of experimental compounds has been aided by the existence of murine cytomegalovirus (MCMV), which is closely related to HCMV. Candidate drug molecules which usually, but not always, possess activity against both HCMV and MCMV, can be investigated for efficacy in this mouse model.

Nucleosides

Unlike the herpes viruses so far discussed, CMV does not encode a thymidine kinase enzyme, which means that the majority of anti-HSV and anti-VZV nucleosides cannot be phosphorylated in the same way, and often prove to be inactive. However, the CMV gene product, UL-97 appears to be able to circumvent this problem in some cases, by acting as a kinase; it appears to be the agent which promotes the phosphorylation of ganciclovir. Despite its poor selectivity profile, which is probably due to host kinase mediated phosphorylation occurring too, ganciclovir (Figure 6.3) is used clinically against CMV due to the severity of the infection. Foscarnet (Figure 5.25), which also has a poor therapeutic index, is also used clinically, especially in cases where there is evidence of ganciclovir resistance emerging.

The research into phosphonate analogues of monophosphates has produced a number of candidates for potential CMV chemotherapy such as HPMPC and SR3727. Such molecules circumvent the need for the first kinase activation. Selectivity in such compounds is achieved as a consequence of the differing effects of the phosphonate–diphosphate (equivalent to the triphosphate) on the viral and host polymerases. Pharmacokinetic and other problems with the phosphonates have been reduced with prodrug formulations: cHPMPC and SR3727A are phosphonate ester prodrugs (Figure 6.9) which are metabolized to their parent forms, prior to further phosphorylation.

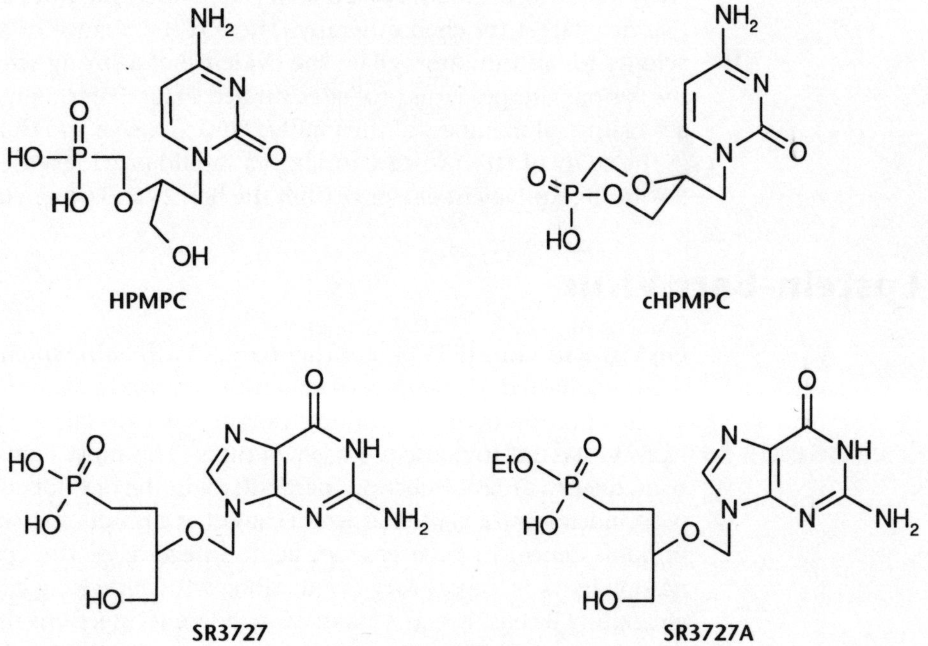

Figure 6.9 The anti-CMV phosphonates HPMPC [9-(3-hydroxy-2-phosphonylmethoxypropyl) cytosine] and SR3737 with their respective prodrugs, cyclic HPMPC and SR3727A.

Non-nucleoside leads

A series of compounds resembling nucleosides, but based on glycosylated benzimidazoles such as BDCRB (Figure 6.10), has emerged as a potent class of anti-CMV compounds. Despite their structure, these molecules do not act as polymerase inhibitors in their triphosphate forms; they are believed to act against some process in late stage viral maturation. Extensive compound screening has also produced examples of many non-nucleosides with anti-CMV activity, some of which have been looked at in the murine model, with mixed success. The mode of action of such molecules is often unknown, but is often the source of new information about biochemical targets.

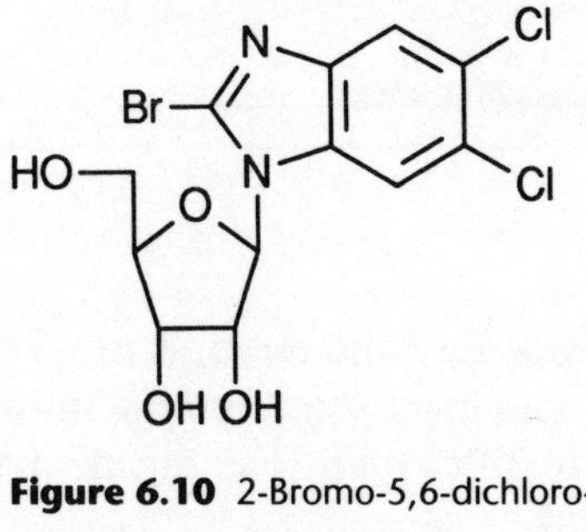

Figure 6.10 2-Bromo-5,6-dichloro-1-(β-D-ribofuranosyl) benzimidazole, BDCRB.

HCMV Protease

Cloning and expression of the HCMV serine protease has allowed screening to commence for inhibitors of this enzyme, which provides a particularly attractive new target for chemotherapy. There is the chance of a broad spectrum of activity for an inhibitor, given the evidence of a strong sequence homology in the various herpes virus proteases studied so far. Potentially, a single drug could act against all members of the family. The extensive and fruitful efforts invested in the study of HIV protease inhibitors should provide a useful paradigm to follow in the equivalent enzymes from the herpes and other virus families.

Epstein–Barr Virus

Epstein–Barr virus (EBV) is another herpes virus with which virtually all people become infected, through contact with body fluids, at some time in their lives. When infection occurs in young children, with sub-clinical symptoms, immunity is conferred to subsequent encounters. The most commonly encountered consequence of EBV infection, particularly in the developed world, is infectious mononucleosis (or glandular fever) caught as a result of kissing or sexual activity in adolescence. In each case an acute infection of the epithelial cells of the nasopharynx is the primary event, along with a life-long latent infection of the circulating B cells. It is as a result of such latent infections that the more sinister consequences of EBV are manifested; the virus is also implicated in a range of other conditions both as the causative agent or by association (Table 6.3). The mechanisms of many of these conditions and the exact roles of EBV are poorly understood, but they are of clear clinical significance.

Table 6.3 Conditions in which Epstein–Barr virus is implicated

Affected tissues/Disease	Role of EBV
Lymphoid	
Burkitt's lymphoma	Cofactor, non-essential
B cell immunoblastic lymphomas	Causative agent
T cell lymphomas	Clonal association
Hodgkin's disease	Clonal association
Epithelial	
Nasopharyngeal carcinoma	Essential factor
Parotid carcinoma	Clonal association
Hairy leukoplakia	Causative agent
Lymphoid and Epithelial	
Infectious mononucleosis	Causative agent

Pathogenesis, latency

There are three distinct phases of EBV infection, acute (potentially symptom-producing), latent (dormancy) and reactivated (producing associated malignancies). The pathogenesis of acute EBV conditions can be manifested directly by cell cytolysis, or can be the consequence of an over-reaction of the immune system to viral antigens. The 178–190 kb linear double-stranded DNA genome of EBV can be circularized to form the EBV-episome, which is the state in which the virus lies dormant. Infectivity depends on the way in which the genome is replicated: the circularized DNA episome can be copied by cellular DNA polymerases, whilst the virus-productive linear form is produced under the control of a virally encoded enzyme. The reactivated state, when EBV is implicated in various tumourogenic conditions, is the least well understood, although many co-factors are implicated, notably the immunosuppression caused by malaria resulting in Burkitt's lymphoma.

Chemotherapy

Cellular assays exist for EBV and many compounds have been screened. Individual compounds from each of the classes described below inhibit only the productive infection stage of the virus, so these may have a use in the treatment of infectious mononucleosis, but probably not other EBV-associated conditions. Despite the lack of a thymidine kinase, the active compounds claimed include some, but by no means all of the active nucleosides against other contemporary herpes viruses, such as acyclovir. This perhaps suggests some other gene product which might promote phosphorylation. Pyrophosphate analogues are also effective, and, unlike all other herpes viruses, the replication of EBV is also inhibited by AZT. However, such conventional forms of chemotherapy would have no effect on the latent infection or cell immortalization, which could perhaps be a future target for nucleic acid based therapeutic approaches.

Other human herpes viruses

There are other known herpes viruses which infect humans, such as human herpes Virus-6 and HHV-7, although the consequences of such infections are poorly understood. Interest in HHV-6, which is closely related to CMV, has been focused on its possible implications in the course of HIV infection and some evidence of links with multiple sclerosis. Cellular assays are available for this virus and inhibitory molecules are starting to be identified; unsurprisingly, these are often the inhibitors of other herpes viruses.

7 Human immunodeficiency virus and AIDS

Introduction

In 1981 a syndrome was first recognized which showed symptoms more usually ascribable to patients whose immune systems had a clinical basis for being compromised. The condition became known as **acquired immune deficiency syndrome** (**AIDS**), and the aetiological agent was identified as a virus soon afterwards; this facilitated research on an unprecedented scale in the search for possible treatments. Ever since, **human immunodeficiency virus** (**HIV**) has continued to make headlines, both through the spread and notoriety of the disease and controversies over its control and treatment.

Table 7.1 Members of retrovirus family, Retroviridae.

Subfamily	Virus	Disease
Oncovirinae	Human T-Cell Leukaemia virus (HTLV-I)	T-cell leukaemia.
	Human T-Cell Leukaemia virus (HTLV-II)	Hairy cell leukaemia
Lentivirinae	Human immunodeficiency virus (HIV 1)	Immune deficiency
	Human immunodeficiency virus (HIV 2)	Immune deficiency
	Simian Immunodeficiency virus (SIV)	Monkey immune deficiency

HIV is an example of a **retrovirus** (see Table 7.1). These are RNA viruses which replicate by transcribing their genetic information from RNA to DNA, inserting this DNA into the host genome (pro-viral DNA), and ultimately forming new RNA from the pro-viral DNA template. This RNA to DNA **reverse transcription** is brought about by an RNA dependent DNA polymerase, otherwise known as **reverse transcriptase** (**RT**), encoded by the virus. The closely related HIV viruses, HIV-1 and HIV-2, are part of a sub-family termed **lentiviruses**, a name derived from the slow (Latin, *lente*) onset of the disease. The genome of these relatively simple viruses is made up of two molecules of single-stranded positive sense RNA with a molecular weight of about 3,000 kDa. The morphological structure of the virions includes both structural and functional proteins within an enveloped 100 nm diameter particle. The gene products include a **protease** enzyme and various regulatory factors, but no nucleoside kinase enzyme is encoded. The genomic maps of HIV are shown in Figure 7.1.

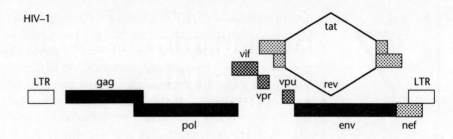

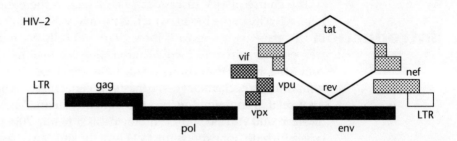

Figure 7.1(a) The genomic maps of HIV-1 and HIV-2.

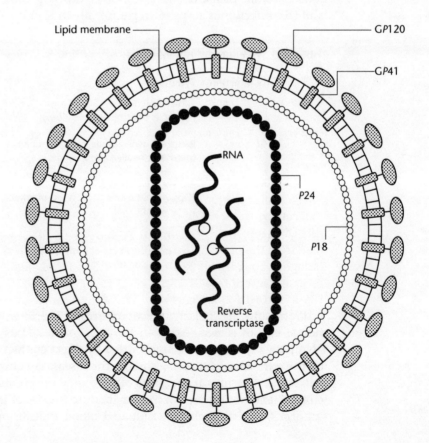

Figure 7.1(b) Structure of the HIV viron.

The primary sites of HIV infection are the helper T-lymphocytes, vital links in the immune system, which are specifically targeted through interaction of their membrane **CD-4 protein** and the HIV surface protein **gp 120**. Viral replication begins very quickly; the immune system reacts to this challenge, but is eventually swamped by the infection and gradually becomes deficient due to the fall in number of CD4+ T-cells. At this stage, the individual becomes highly vulnerable to infection by other viruses and microbial organisms (or **opportunistic infections**) which result in morbidity and eventually death due to more prolonged and severe symptoms, because there is no effective immunological defence against the invading organisms.

The course of HIV disease can be related to the concentration of the CD4+ helper T cells in the blood, which is normally 800–1200 cells per mm^3 of blood. In HIV patients, those with more than 500 cells per mm^3 are defined as being HIV positive, these individuals are otherwise healthy despite the sub-clinical condition, but are infectious. As the CD4 level drops the lymph nodes swell, and symptoms are classified as persistent generalized lymphadenopathy (PGL); AIDS-related complex (ARC) is subsequently defined in those with CD4 counts between 200 and 500 and full blown AIDS below 200. Ultimately, HIV infects cells which do not express the CD4 protein and cause disease in the muscles and central nervous system, manifested by muscular wastage and AIDS dementia. An overview of the course of HIV infection is shown in Table 7.2. The natural history of HIV infection is also shown graphically in Figure 7.2

Table 7.2 The course of HIV infection

HIV Infection	Symptoms	CD4 count
	Sub-clinical disease: (Person is healthy, but infectious)	~normal
PGL	Persistent generalized lymphadenopathy (persistent swollen lymph nodes)	>500
ARC	AIDS-related complex (Opportunistic infections, fever, diarrhoea, some weight loss)	200–500
AIDS	*Mild*: Constant infection and/or neoplasms *Severe*: Severe infections, myopathy, nervous system disease (dementia).	<200

HIV is not a particularly infectious virus; it is primarily carried in the blood, though also in semen, either in the form of free virus or within an infected T-lymphocyte. Infection is transmitted by direct contact of body fluids, including through sexual contact, although the disease is better transmitted by males than females. Lesions arising from other genital infections, which increase blood to blood (or semen to blood) contact increase the risk of infection. Transmission can also occur through contaminated blood transfusions, blood products or

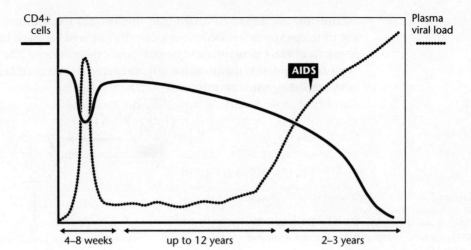

Figure 7.2 Plot of CD4+ count and viral load vs. time.

shared hypodermic needles. The virus can also be carried across the placenta, though only about 20% of babies born to HIV positive mothers become infected.

Chemotherapy of AIDS

An ever-increasing number of agents have reached clinical evaluation for chemotherapeutic intervention against HIV. Set against an increasing under-standing of the complex mechanism of the infection, objectives can be set for desirable effects with such putative treatments. The fundamental aim of chemo-therapy is to prevent viral replication, thus reducing the viral load and delaying symptoms, which ultimately prolongs survival (Figure 7.3). By reducing the viral replication rates the breakthrough of new phenotypes is also slowed, although paradoxically the selection pressure of an antiviral agent increases the evolution rate of resistant phenotypes. The chances of clearing the infection

Figure 7.3 Chemotherapeutic objectives of HIV chemotherapy: interrupting the events of the lowest levels of the triangle produces effects on those above.

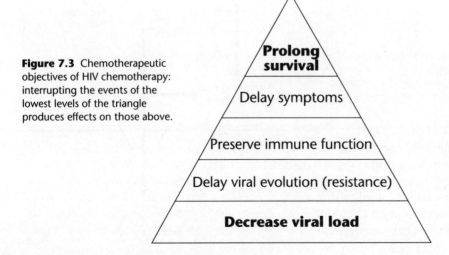

completely are probably unrealistic; this would require the elimination of all cells carrying pro-viral DNA, but clear clinical benefits have been demonstrated through chemotherapy, most notably in combinations. The effectiveness of a drug or drug combination is usually measured in terms of changes in the CD4 and circulating virus counts.

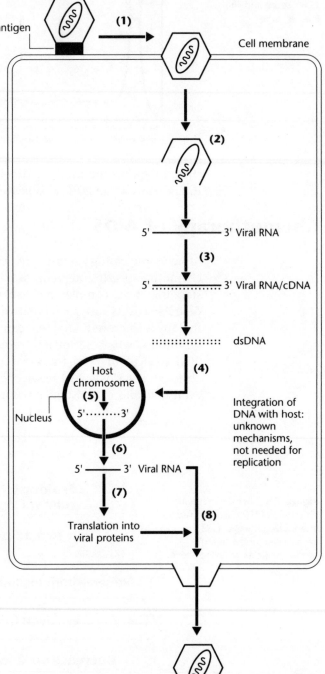

Figure 7.4 The HIV replicative cycle. The stages of replication are:
(1) The virus attaches to the cell through gp120 CD4+ receptor interaction and the viral envelope fuses to the cell membrane.
(2) Uncoating of the virion: the nucleocapsid complex and viral enzymes are released.
(3) Transcription of viral RNA into DNA, catalysed by reverse transcriptase; RNase H digestion of RNA, then double-stranded DNA formation.
(4) Incorporation of dsDNA into the host genome. This reaction is catalysed by viral integrase enzyme.
(5) Transcription of pro-viral DNA.
(6) Transcription of pro-viral DNA to viral mRNA.
(7) Translation of mRNA into viral proteins.
(8) Post-transitional modifications, processing and maturation of new virions prior to release, such as cleavage of polypeptides to functional proteins by protease enzyme, and trimming of sugars by glycosidases.

A wealth of information on the molecular biology of HIV has resulted from the extensive scientific studies into the disease, facilitating the development of biochemical assays for specific inhibitory molecules. This section introduces these targets for chemotherapy and describes agents which have reproducible *in vitro* activity against them. An ever-increasing number have progressed to clinical trials and the marketplace, with demonstrable beneficial effects. However, there is a significant problem of **resistance** developing to the agents, which is discussed in the final section.

HIV Replicative cycle and targets for chemotherapy

In the replicative cycle of HIV there are several well-defined mechanisms which are targets for chemotherapy, as highlighted in Figure 7.4. The extent to which individual targets have been studied is somewhat reflected in the arrangement and detail within the following sections.

HIV reverse transcriptase inhibitors

The RNA-dependent DNA polymerases, or reverse transcriptases are unique to retroviruses and consequently constitute attractive targets for chemotherapy. HIV reverse transcriptase (HIV-RT) is an essential component of the virion and has multi-functional roles. The enzyme has been cloned and expressed, so that amounts of protein have long been available for RT inhibition assays. Latterly, crystallographic studies have provided substantial insight into the structure and function of this heterodimeric enzyme (Figure 7.5). Its two constituent polypeptides are 51 kDa and 66 kDa in size; the so-called p51 subunit is a cleavage product of the p66 protein, lacking the RNase functional region. Molecules which inhibit RT can be divided into nucleoside and non-nucleoside (NNRT) inhibitors. Currently licensed therapies are limited to the former class, whilst candidates are under clinical evaluation from both.

Nucleoside analogues as reverse transcriptase inhibitors

HIV does not encode a kinase, so any potential nucleoside therapeutic agent must be a substrate for the host kinases, in order to produce the active triphosphate form of the drug. Given the high degree of specificity of host kinases, nucleosides with anti-HIV activity have differed little in structure from the natural substrates. A consequence of this is that selectivity is harder to achieve, thus the therapeutic indices are often somewhat lower than with corresponding anti-herpes drugs. Many of the active anti-HIV molecules are analogues of the 2'-deoxy nucleoside, thymidine, the triphosphate of which is one of the natural substrates for RT. Some examples are shown in Figure 7.6. The most notable difference in each case is the absence of a 3'-hydroxyl group, which makes the compounds mandatory chain terminators of the nucleic acid polymerization process. Oxathiolanes, 3TC and FTC, have the most obviously different structures; not only are there two heteroatoms in the ring, but also they possess the

(a)

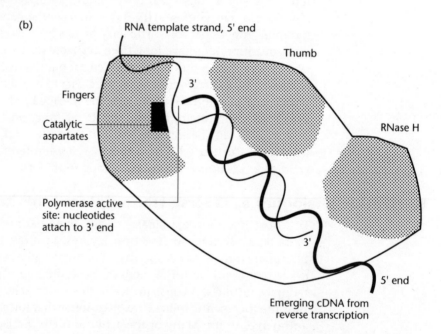

Figure 7.5 Schematic representations of the structure of HIV-1 RT showing the general structural features and DNA–RNA hybrid binding. A clearer model of the 3-dimensional structure of RT can be perceived if a structure is imagined based on a right hand (diagram A). The substrate cleft is between the thumb and fingers, with three aspartate residues on the lower fingers, which bind magnesium to catalyse the addition of complementary DNA nucleotides to the RNA template spiralling across the palm like an extended spring (diagram B). The palm is the site of a hydrophilic pocket in which some inhibitors can bind. Above the wrist is the RNase H region, which may act in a concerted fashion with the rest of the enzyme in the cleavage of viral RNA from the reverse transcribed cDNA.

(b)

unnatural or L- absolute configuration; the enantiomeric forms of these molecules have similar activities, but are less selective.

A number of phosphonates have been shown to be active against HIV, the use of such a phosphate isostere or monophosphate prodrug strategy has been particularly targeted at HIV therapy. PMEA, in the form of its prodrug bis-POM PMEA (Figure 7.7) continues to undergo clinical evaluation.

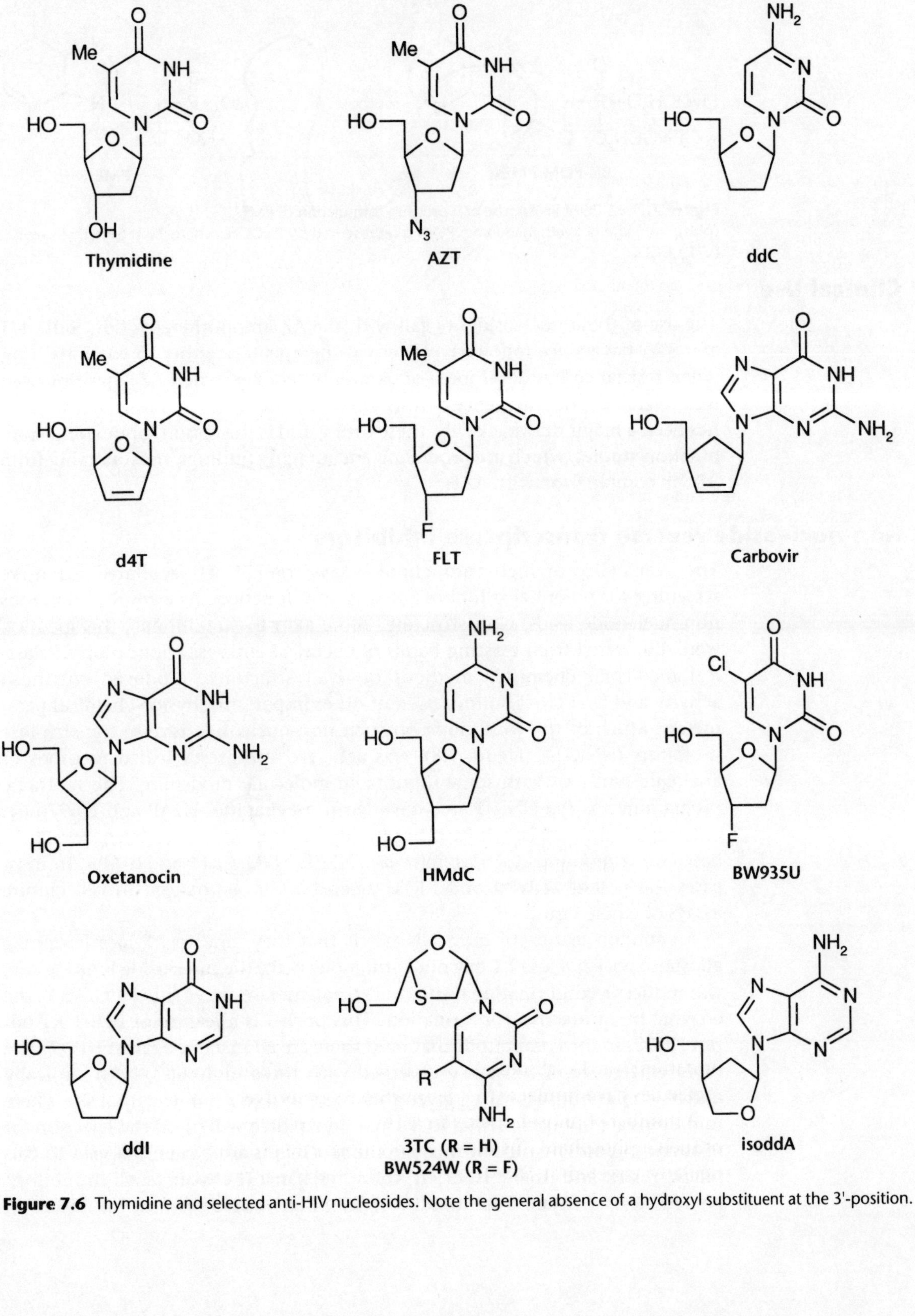

Figure 7.6 Thymidine and selected anti-HIV nucleosides. Note the general absence of a hydroxyl substituent at the 3'-position.

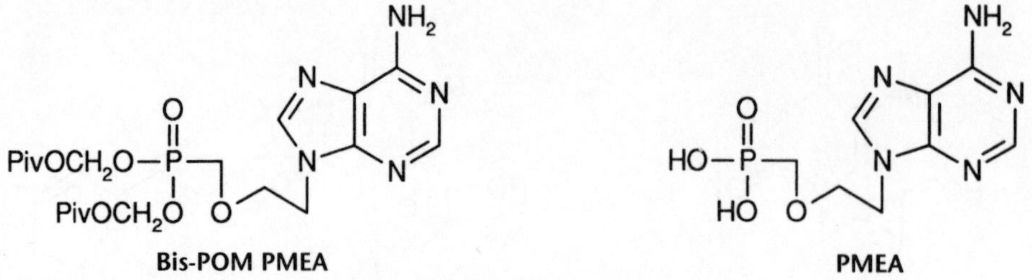

Figure 7.7 Bis-POM PMEA, the oral prodrug formulation of PMEA (phosphonylmethoxyethyl)adenine. POM is represented by PivOCH$_2$ where Piv is trimethylacetyl, (CH$_3$)$_3$CCO.

Clinical Use

The use of these nucleosides began with the AZT monotherapy; ddC, ddI, d4T and 3TC have since followed as licensed therapeutics, either in combinations, when resistance has developed, or as monotherapies, when AZT has not been well tolerated. The emergence of resistant and sometimes cross-resistant mutants has been a major drawback with these drugs, and is the reason for ongoing combination studies, which are producing encouraging findings, most notably from 3TC in combination with AZT.

Non-nucleoside reverse transcriptase inhibitors

The availability of high throughput assays for HIV-RT facilitated extensive screening for potential inhibitors of this viral function. As a result, numerous non-nucleoside leads, with structures more akin to "traditional" therapeutics, were discovered from existing banks of chemical entities. Application of classical medicinal chemistry methods on lead structures, produced enhanced activity and facilitated optimization of other important physico-chemical parameters. Much of the work done on such non-nucleoside reverse transcriptase inhibitors (NNRTIs) (Figure 7.8), was achieved using established methods of analogue synthesis with some input from molecular modelling. The synthetic programmes in the NNRTI area have led to nevirapine, BHAP and L-697-661, entering clinical trials, whilst other highly potent series have been developed based on compounds such as, *inter alia*, HEPT, TIBO, PETT and α-APA. In most cases the potent activity of NNRTIs against RT is reproduced in cell culture assays of whole virus.

A common feature of most NNRTIs is that they bind to RT in a lipophilic allosteric pocket, close to, but not contiguous with, the nucleoside binding site; this induces a conformational shift in the polymerase active site which locks the enzyme in an inactive conformation. This pocket is a feature of HIV-1 RT but not HIV-2, so the compounds that bind there are all inactive against HIV-2. The rapid emergence of resistance observed with these molecules, both clinically and *in vitro*, is not surprising, given that they bind to a non-essential site where mutations are not detrimental to RT function, which will speed the selection for resistant phenotypes. In most cases, the mutations have been mapped to this binding site and there is often cross resistance between such inhibitors. Nonetheless, a detailed knowledge of RT point mutations and inhibition of

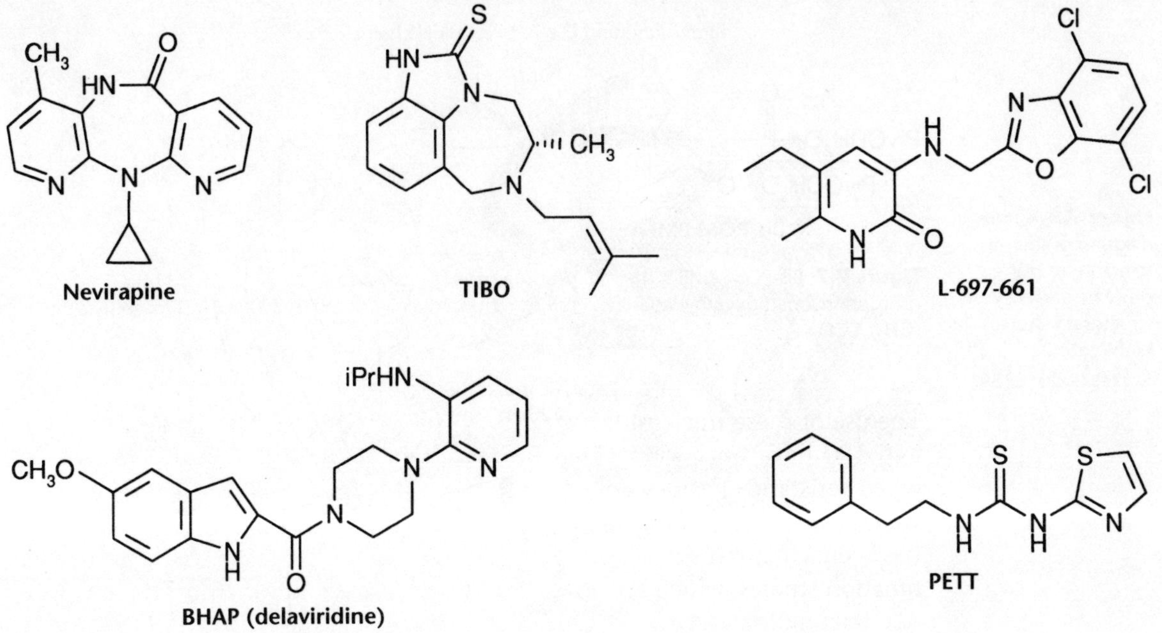

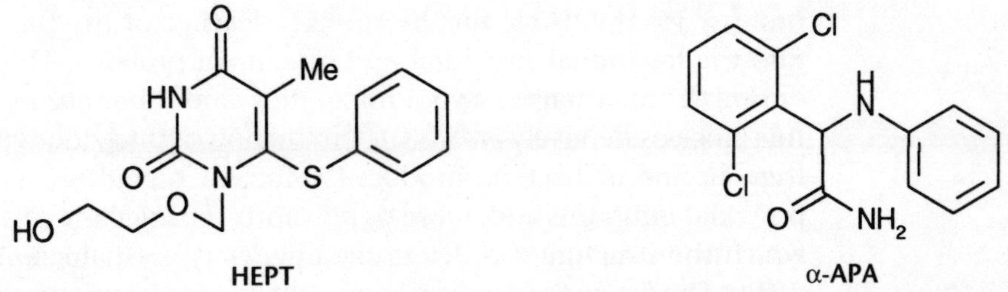

Figure 7.8 Non-nucleoside HIV-RT inhibitors.

Structures labelled: Nevirapine, TIBO, L-697-661, BHAP (delaviridine), PETT, HEPT, α-APA.

mutant RTs does suggest some scope for the continued clinical use of NNRTIs in combination therapies. This occurs when the NNRTI-resistant phenotype may either become hypersensitive to a second drug or the resistance to a second drug is delayed. More recently, the availability of X-ray crystallographic stuctures of such inhibitors, co-crystallized with RT, has produced further insight into their action. The bound NNRTI's have indeed been found in the same allosteric-pocket, where the binding interactions are consistent with mapped mutations introducing both resistance and cross resistance within the series. Further synthetic work based on such X-ray data will lead to new, more rationally designed, NNRTIs in the future, particularly those which are desigend to target the nucleotide binding site in the enzyme.

HIV Protease inhibitors

Amongst the functional proteins in the HIV genome which provide a target for chemotherapy, the HIV protease enzyme has come a close second to RT in terms of efforts towards its inhibition. This homodimeric, C_2-symmetric, aspartyl

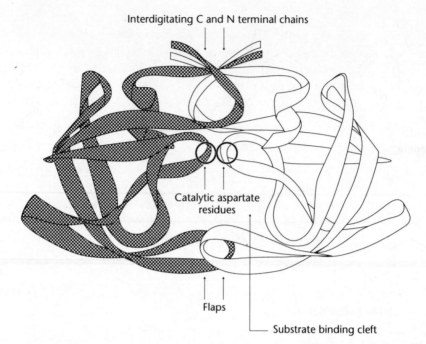

Figure 7.9 Ribbon diagram of the structure of HIV protease with key structural features highlighted.

Interdigitating C and N terminal chains

Catalytic aspartate residues

Flaps

Substrate binding cleft

protease is responsible for the specific cleavage of the translated polypeptides into the individual structural and functional proteins. This is most important during the final stages of viral maturation and it has been shown that blocking this process effectively arrests the production of infectious virions. Expression of this enzyme in bacteria produced sufficient quantities to develop assays for potential inhibitors and, more significantly, enabled crystals to be formed, from which the structure was determined by X-ray crystallography (Figures 7.9 and 7.10). The enzyme has also been chemically synthesized, the folding and

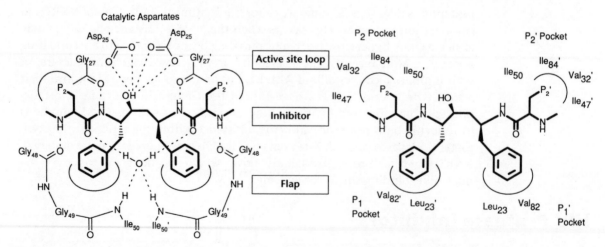

Figure 7.10 Simplified 1-dimensional representations of the binding of a typical inhibitor core in the active site of HIV protease, showing hydrogen bonding patterns and hydrophobic pockets. Refinements of this simplistic model have provided the basis for much of the molecular design.

structure of the synthetic version being identical to the naturally derived molecule. HIV protease is perhaps the most extensively studied enzyme to date from a medicinal chemist's point of view. This has included the solving of structures containing co-crystallized inhibitors, leading to much computer aided design and consequent synthetic activity, which has led to better and more efficacious inhibitors. Structure elucidation was achieved much more rapidly for protease than for RT.

Synthetic design

The design of HIV protease inhibitors had something of a head start from ongoing work on other aspartyl proteases, such as renin, and useful models were constructed using the specific cleavage of Phe–Pro (Figure 7.13), which is unique to the HIV enzyme. A key feature of rationally designed inhibitors has been a mimic of the tetrahedral intermediate formed during cleavage of the scissile amide bond of the substrate; this can hydrogen bond to the catalytic aspartate residues in the active site (Figure 7.11). Such non-scissile linkages (Figure 7.12)

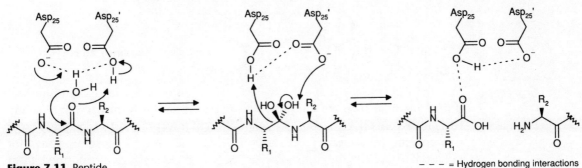

Figure 7.11 Peptide cleavage, by the generally accepted mechanism, as catalysed by the twin aspartate residues in the HIV protease enzyme.

– – – – = Hydrogen bonding interactions

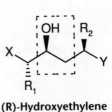

Figure 7.12 Natural amide linkage and some examples of non-cleavable isosteres. The hydroxyl residues are important, given that they will generally be within hydrogen bonding distance of both aspartate carboxyls.

Amide

(S)-Hydroxyethylene

(R)-Hydroxyethylene

Amine

Statine

Dihydroxyethylene

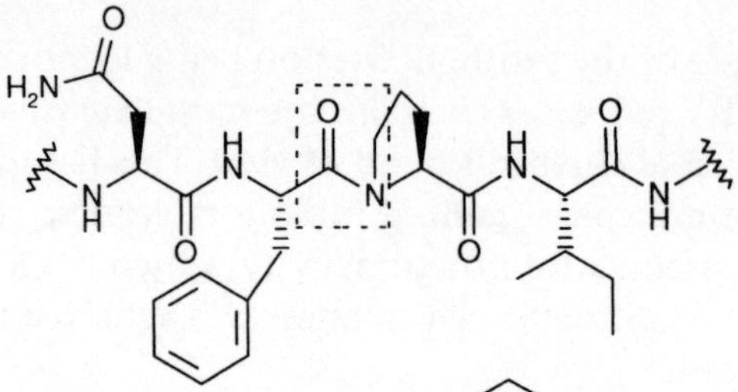

Figure 7.13
A recognized cleavage sequence unique to HIV protease (Asn-Phe-Pro-Ile) and selected HIV protease inhibitors based on this, which exemplify the pattern of diminishing peptide character within the molecules.

Recognized cleavage sequence of HIV protease

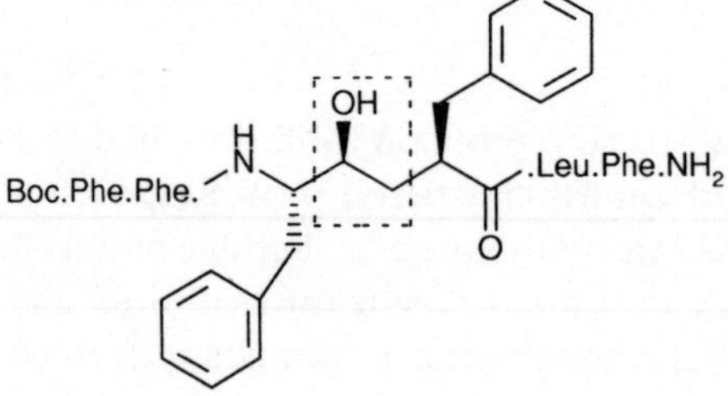

L-364,505

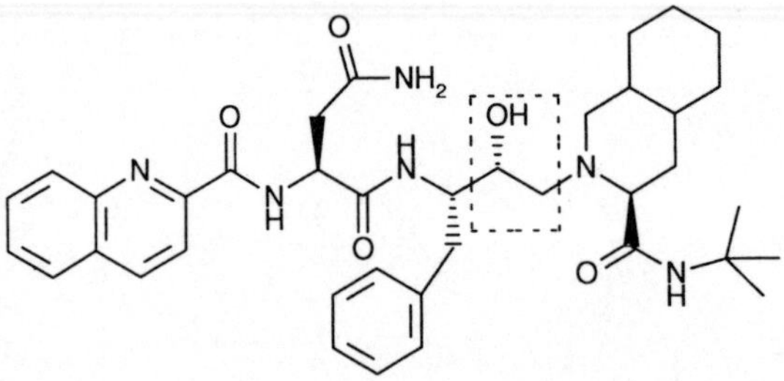

Saquinavir, Ro 31-8959

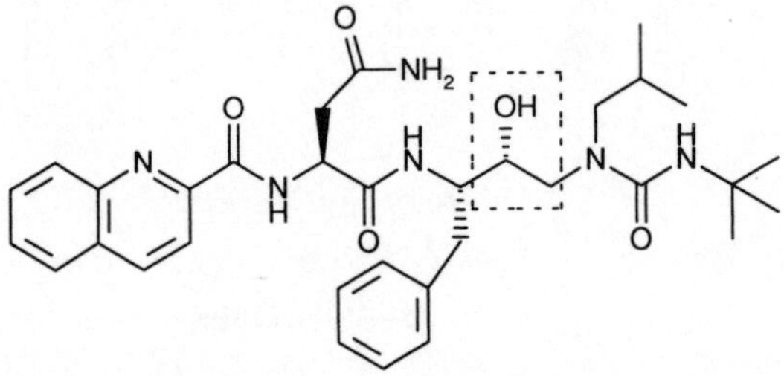

SC 52151

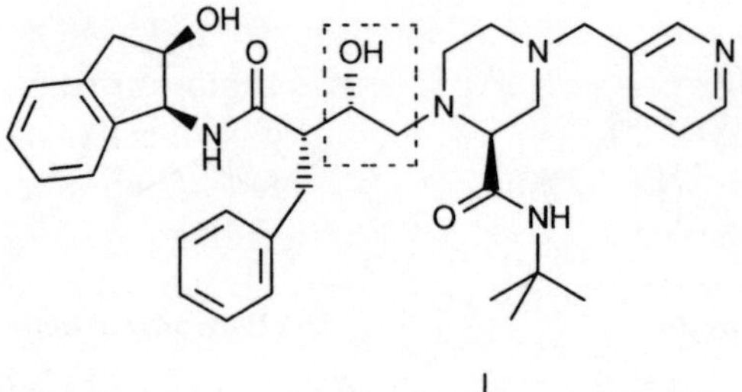

Indinavir, L-735,524

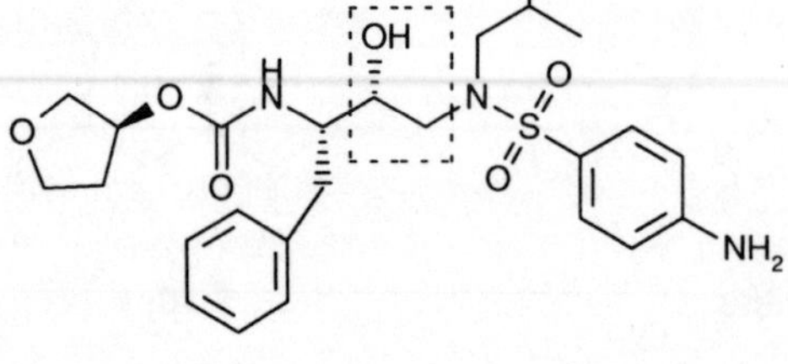

VX 478

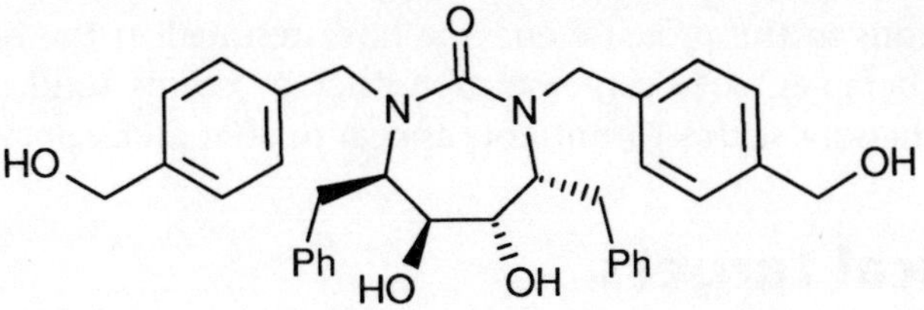

Figure 7.14
Examples of non-peptidic structures, which had a rational basis in computer modelling with a knowledge of X-ray structures.

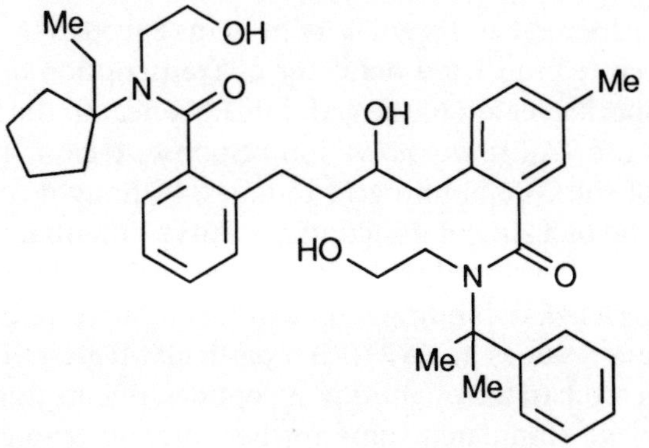

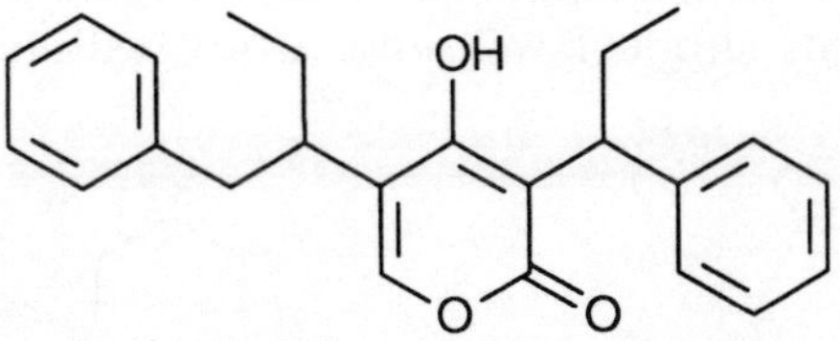

XM323

AG-1284

U96988

have often been the central part of otherwise C_2-symmetric inhibitors, which recognized this structural feature of the enzyme. Many early inhibitors were simply short peptide side chains linked through nitrogens at the two ends of a non-scissile linkage. Subsequently, design work and further synthesis introduced substituents which mimic natural substrate side chains, whilst retaining or improving activity. This approach, followed by many organizations, has produced a large number of powerful inhibitors (sub-nanomolar inhibitory concentrations are common), which have been active *in vitro* against both the enzyme and virus. The peptoid examples chosen here are used to emphasize diminishing peptide character in the molecules (Figure 7.13); other examples are listed in Appendix 2. Extensive screening and modelling studies has also led to classes of potent inhibitors with no peptidic character at all (Figure 7.14).

The clinical promise of most of these compounds has been limited, despite their undoubted potency. Many of the exploratory molecules were unsatisfactory due to poor stability and/or pharmacokinetics, factors which were improved as a result of specific chemical manipulations on the structures. Some 20 examples have reached some form of clinical evaluation, normally in combination with an RT inhibitor. The recent approval of saquinavir is the first of what appears to be a number of candidates for commercial manufacture. However, mutations in the protease enzyme have resulted in the onset of resistance to such therapies, so a protease monotherapy seems unlikely, though multiple combinations with RT inhibitors appear to offer much potential.

Other biochemical targets

HIV TAT Inhibitors

Amongst the structural and regulatory proteins encoded by HIV is a *trans*-acting polypeptide termed the **Trans-Activator of Transcription** or **TAT**, which acts by binding to a specific region of the genomic RNA near to the long terminal repeat (LTR) termed the TAR (*trans*-activation response) region. The action of TAT, a polypeptide of some 86 amino acid residues, is thought to promote viral RNA synthesis, so the blocking of its action presents a potential therapeutic target. *In vitro* models were developed which were sensitive to specific inhibition of TAT function. The TAT–TAR binding interaction is thought to be due to the interaction of arginine residues in TAT with a particular region of TAR RNA; anti-TAT activity was demonstrated by short polypeptides, rich in arginine residues, molecules which have stimulated some further interest. Systematic screening programmes and subsequent synthesis, produced one series of putative TAT inhibitors. The benzodiazepine-based Ro 24-7429 (Figure 7.15) briefly underwent clinical trials, although was subsequently withdrawn.

Figure 7.15 Chemical structures of Ro 24-7429 and JM-3100.

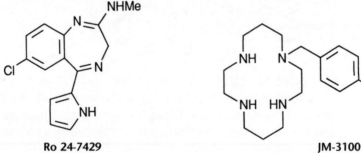

Ro 24-7429

JM-3100

Gp120-CD4 interaction

The specific interaction between the HIV surface protein gp120 and the CD4 surface receptor complex on the T-lymphocytes has also been explored as a potential therapeutic target. In theory, by binding recombinant CD4 or a synthetic surrogate to virions in the serum, the crucial cellar recognition due to gp120 could be interfered with and infection prevented. The drawbacks of such an approach include the difficulties in maintaining sufficient levels of the drug and the large number of potential receptors on each of large numbers of individual HIV virions. Sulphated polysaccharides and other relatively small synthetic molecules, have been investigated in addition to biological macromolecules as potential inhibitors of this interaction and several have undergone preliminary clinical trials.

Uncoating of the virus

The uncoating of a virus in a newly infected cell is a key step in the viral reproduction cycle, and is a proven target for some drug molecules, such as the lipophilic amines used to treat influenza. A series of compounds based on the bridged polyamine JM-3100 (Figure 7.15) is believed to act in this fashion against HIV.

Glycosidase inhibitors

A key step in the maturation process of a virus is the clipping of saccharide linkages in glycoprotein structures attached to the virion. This process has been inhibited by a number of amino sugars, which show specificity akin to the absolute configuration of their attached hydroxyls, thus castanospermine and deoxynojirimycin (DNJ) are known glucosidase inhibitors. These molecules have also been shown to inhibit HIV replication, although they do not specifically inhibit a virally encoded target; the glucosidase is a host enzyme. The problematical low lipophilicity of DNJ was increased by the formation of N-butyl-DNJ, which has been evaluated further for its anti-HIV potential (Figure 7.16).

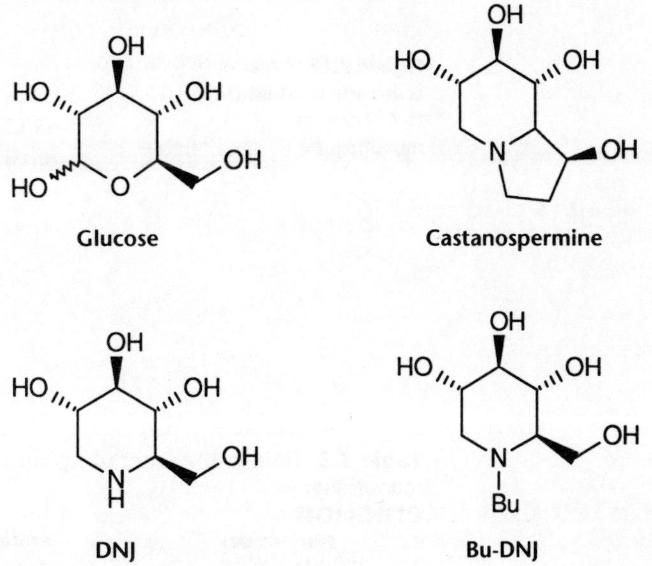

Figure 7.16 Glucose and glucosidase inhibitors castanospermine, deoxynojirimycin and its butylated analogue.

Glucose

Castanospermine

DNJ

Bu-DNJ

Integrase inhibitors

The HIV integrase enzyme, a 32 kDa protein encoded on the *pol* gene, catalyses the integration of retroviral DNA into host DNA. This enzyme has no functional counterpart in human cells, so its inhibition presents an attractive chemotherapeutic target. Early work has produced some potent inhibitors, but these have either lacked specificity or had undesirable physical properties. The recent developments of high throughput *in vitro* assays and structure elucidation of the catalytic domain of integrase, give the area much future potential.

Resistance

One problem in the chemotherapy of HIV-related conditions is that the level of a drug's antiviral activity is not maintained with continued use (Figure 7.17). The prime factor which introduces this lack of sustainable activity is the readiness by which HIV can mutate its genetic make up. The HIV reverse transcriptase enzyme produces new viral RNA with a relatively high level of miscopies: thus, coupled with rapid reproduction, the evolutionary processes are rapidly amplified. Consequently, specific mutations can produce drug-resistant phenotypes very quickly (Table 7.3). This is manifested in the changes in the clinical effectiveness of therapies. Changes in such clinically isolated viruses can be mapped to specific amino acid mutations (Table 7.4) in the functional proteins, using the techniques of molecular biology. Passaging the drug *in vitro* against the virus will induce the same mutations; the rate at which these develop, in terms of number of passages required, reflects the time taken for clinical effects to emerge.

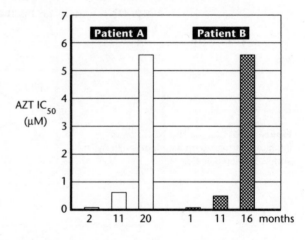

Figure 7.17 Onset of resistance: HIV sensitivity to AZT during monotherapy.

Table 7.3 Time to the onset of significant resistance during monotherapy

Nucleosides		Protease inhibitors	
AZT	6-12 months	L-735,524	3-6 months
ddI	6-12 months		
ddC	~ 12 months		
NNRTIs	days-weeks		
3TC	days-weeks		

The rate of onset of different mutations and their significance in terms of the relative effectiveness of each drug does vary. This can be the consequence of single, double or more amino acid changes. It will certainly be a measure of the importance of those mutations on the effectiveness of the enzyme to perform its

Table 7.4 Mutations arising in HIV-RT as a result of exposure to AZT

Amino acid/Position		Mutant	Increase of AZT IC_{50}
Met 41	⟶	Leu	4x
Asp 67	⟶	Asn	nil
Lys 70	⟶	Arg	8x
Thr 215	⟶	Phe/Tyr	16x
Lys 219	⟶	Gln	nil

required function. Some mutations are manifested quickly, but have relatively little impact on the effectiveness of the drug, such as some of those arising during AZT therapy (Figure 7.18). Others might be selected for very quickly and have a profound impact on the sensitivity of the mutant enzyme to the drug. For this reason 3TC has no potential as a monotherapy due to the valine-184 mutation effectively rendering the drug redundant. However, this mutant form is hypersensitive to AZT, hence the effectiveness of the combination.

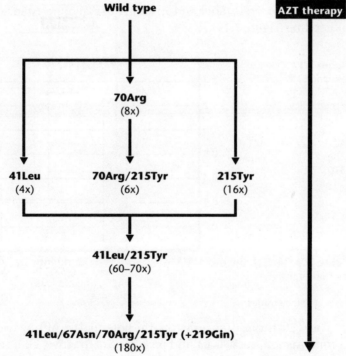

Figure 7.18 Evolution of AZT resistance: multiple mutations and increase in IC_{50}.

Combination therapy

With a knowledge of the specific patterns of mutations (Table 7.5) rational approaches can be made to combination therapies. Obviously, if a phenotype is cross resistant to two particular drugs, then a combination of the two would have little potential either. However, with a knowledge that resistance to each drug was conferred by separate mutations, then the combination therapy becomes a

Table 7.5 Cross-Resistance Patterns with RT inhibitors

Inhibitor	Important mutations	Cross-Resistant with
AZT	41Leu, 215Tyr etc	3'-azido analogues
ddI	65Arg, 74Val, 184Val	ddC, 3TC
ddC	65Arg, 74Val, 184Val	ddI, 3TC
3TC	184Val	ddI, ddC
d4T	75Thr	ddI, ddC
NNRTIs	181Cys etc	NNRTIs

distinct possibility. Thus a combination of ddC and ddI is of little thera-peutic use when resistant viruses develop due to the same mutations being induced; but the combination of either with AZT does show some benefit, due to different resistance patterns. Much of the knowledge gained in this area has stemmed from *in vitro* experiments, using mutant strains produced in the lab-oratory. Significantly, experiments with clinical isolates of mutant viruses show similar results and the rationally designed combinations are showing clear clinical benefits. Ideally, a combination will show at least additive, if not synergistic, antiviral effects, no cross resistance and an ability to exploit resistance reversals due to the other drug (Figure 7.19).

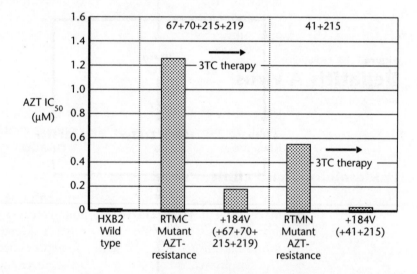

Figure 7.19 Reversal of AZT resistance after passaging the mutant phenotypes with 3TC.

Owing to the resistance factors, the potential for hitting two different targets in a combination, such as RT and protease, is attractive; clinical and *in vitro* results have borne this out. However, other targets also develop resistance to therapy, so subsequently, co-resistant phenotypes which lack sensitivity to either mono-therapy may be evident, thus demanding the need for multiple combinations. Resistance, cross-resistance and co-resistance will always be factors in HIV therapy, but judiciously designed combinations of inhibitors should demon-strate the clearest clinical benefits.

8 Viral hepatitis

The liver is the site of extensive viral replication in many acute infectious conditions, although few of these cause any significant damage to the liver itself. This is due to the site of the viral replication being in the Kuppfer cells of the reticuloendothelial system. True hepatitis viruses, which are drawn from a wide range of viral families, target the hepatocytes; these are grouped together because of their common site of action rather than any virological relationship with each other. There are an increasing number of identified agents; this chapter deals chiefly with the two best defined hepatitis viruses, A and B (HAV, HBV) and introduces agents responsible for some forms of NANB (non-A, non-B) viral hepatitis, HCV, HDV and HEV. Hepatitis viruses are difficult to grow in cell culture, and for this reason only meagre advances in their chemotherapy had been made until recently.

Hepatitis A virus

Hepatitis A virus (HAV) is a member of the *Enteroviridae* family; the relatively small (27 nm) virion has cubic symmetry containing a small single-stranded RNA genome (MW 2,300 kDa) which encodes for only four polypeptides.

Epidemiology and clinical symptoms

The virus, like other enteroviruses, is transmitted by the faecal–oral route; this makes infection particularly common in areas with poor general hygiene standards. A sub-clinical infection is often acquired in childhood which confers future immunity. Epidemics have sometimes been linked to infected shellfish. The symptoms include general malaise, fever and subsequent darkening of the urine and lightening of the faeces, followed by a period of jaundice. The infection is self-limiting and the patient recovers, with a very low incidence of complications and no evidence of latency.

Treatment

Given the small number of gene products, HAV is a poor target for chemotherapy. Prophylaxis is possible, a booster injection of normal human immunoglobulin (HIG) contains enough HAV antibody to prevent an attack for 3 to 6 months. A vaccine has recently been developed, derived from formalin inactivated HAV grown in culture.

Hepatitis B virus

Worldwide, an estimated 200 million people are infected with hepatitis B virus of whom 75% are infected at birth; ultimately, this virus is the single most important cause of cancer in humans. Of the 200,000 new cases each year in the USA 20,000 carriers will emerge, of whom 4,000 will eventually die of cirrhosis and 1,000 of liver cancer.

Hepatitis B virus (HBV) is a member of the family *Hepadnaviridae* (**Hepatitis DNA** viruses) which includes related viruses infecting ducks, ground squirrels and woodchucks. The latter have sometimes been used as *in vivo* models, to aid in compound evaluation; the woodchuck virus is the most closely related to human HBV. As a model, the woodchuck has disadvantages due to the size (requires lots of compound) and hibernation behaviour of the animal. This is being superseded by a recently developed model which uses an HBV-infected human carcinoma cell line grafted into a SCID mouse, which has provided an invaluable investigative tool which utilizes a lot less compound and is acting on the human HBV.

The HBV virion (or Dane particle, after its discoverer) is a 42 nm diameter double shelled structure including a core icosohedral nucleocapsid which contains the genomic DNA (2,200 kDa), an RNA-dependent DNA polymerase and two core antigens. This particle is found in the blood of infected individuals along with two other, non-infectious, structures which are 22 nm diameter spheres or tubules comprised of only the outer layer of the Dane particle, but no nucleic acid.

The genome of HBV consists of a negative strand of DNA attached at the 5'-end to a protein. This is circularized by a variable length, complementary, positive strand of DNA. Despite its small size, comparable to that of HAV, the HBV genome encodes many polypeptides, due to the use of overlapping reading frames. The replication is complex, but a significant feature is the transcription of an RNA intermediate from the completed positive strand DNA; this RNA serves as a template for new negative strand DNA, which is catalyzed by the virally-encoded RNA dependant DNA polymerase. This reverse transcription process (Figure 8.1) is similar to that found in retroviruses, which is a significant feature for potential chemotherapy.

Transmission

HBV is carried only in blood and body fluids, such as semen or vaginal secretions, so sexual activity, shared intravenous needles and contact with other contaminated blood will cause infection; 1 ml of blood can contain up to 10^{10} Dane particles. HBV is much more resilient than HIV, which is transmitted in similar ways, so contaminated media remain potentially infectious for longer.

Clinical aspects.

In about 90% of HBV cases, symptoms similar to those of HAV occur; the patient fully recovers and is immune to further infection. The immune system provides a different response in about 1 in 1,000 cases, and results in fulminant hepatitis,

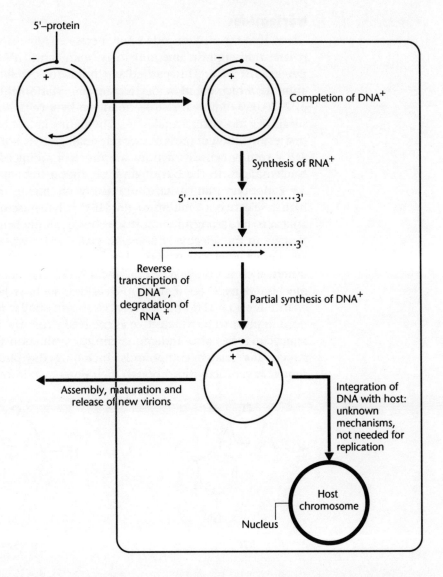

Figure 8.1 Nucleic acid processing in the Hepatiis B virus showing replicative cycle, highlighting the reverse transcription stage.

5'–protein

Completion of DNA$^+$

Synthesis of RNA$^+$

5'.............................3'

.......................3'

Reverse transcription of DNA$^-$, degradation of RNA$^+$

Partial synthesis of DNA$^+$

Assembly, maturation and release of new virions

Integration of DNA with host: unknown mechanisms, not needed for replication

Host chromosome

Nucleus

when abnormally active T-lymphocytes destroy infected hepatocytes and ultimately cause death through hepatic coma. Immunological markers are characteristic of the various forms of chronic HBV infection, the severe form of which kills the individual. Hepatocellular carcinomas may result from integration of the HBV genome into hepatocyte DNA. How this progresses is not fully understood, but many cofactors, such as age and smoking, have been proposed.

Therapies

To date, the virus encoded RNA directed DNA polymerase mentioned above is the only characterized molecular function of HBV that provides a target suitable for drug action, although therapies based on interferons are in clinical use.

Nucleosides

Given that the HBV-encoded polymerase enzyme is similar in function to HIV reverse transcriptase, unsurprisingly, many anti-HIV nucleosides are also effective against HBV. This particularly holds for cytidine analogues, one explanation for which could be due to phosphorylation effects. Relatively high levels of 2'-deoxycytidine kinase are found in liver cells, and this would have a high specificity for cytidine analogues. Thymidine kinase activity, associated with the phosphorylation of most nucleoside analogues, is not so prevalent in these generally resting cells, its actions are more pronounced in active, dividing cells. Nucleosides with the L-configuration appear to be particularly good substrates for 2'-deoxycytidine kinase, ultimately producing triphosphates which show highly selective inhibition of the HBV polymerase over host functions. These appear to be amongst the most selective of all nucleoside agents, comparable in toxicity to the selectively phosphorylated anti-herpes agents such as acyclovir. Of these L-nucleosides, 3TC, has demonstrated clear clinical benefit, whilst others, such as BW524W, L-ddC and 4391W have good *in vivo* profiles in animal models. Many D-configured nucleosides, such as FIAU (Figure 8.3) and the enantiomers of 3TC and 524W have shown similar *in vivo* potential, although none appears to be as selective as the leads from the L-series. The anti-herpetic famcyclovir has also undergone clinical evaluation for HBV and a number of structurally related compounds, including the phosphonate PMEA, display similar *in vitro* activity, although their potency is lower than 3TC.

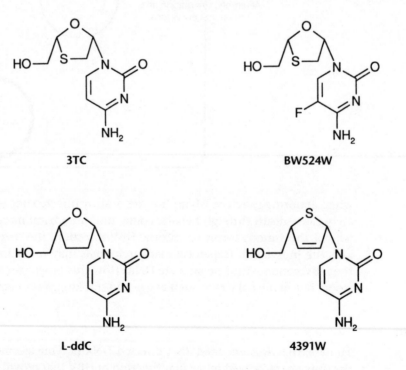

Figure 8.2 3TC, BW524W and related anti-HBV L-nucleosides L-ddC and 4391W.

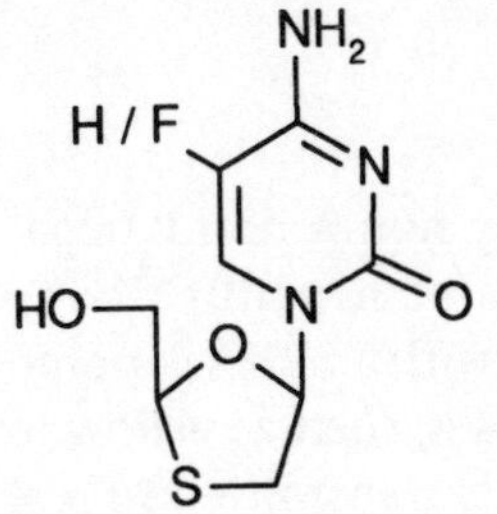

Figure 8.3 Anti-HBV D-configured nucleosides Ent-3TC, BW523W, FIAU and acyclics famiciclovir and PMEA.

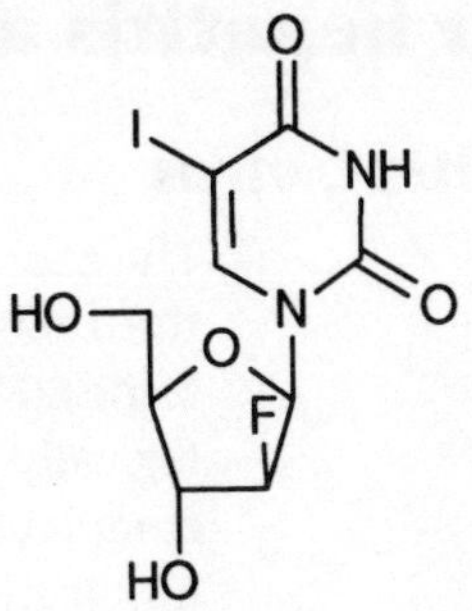

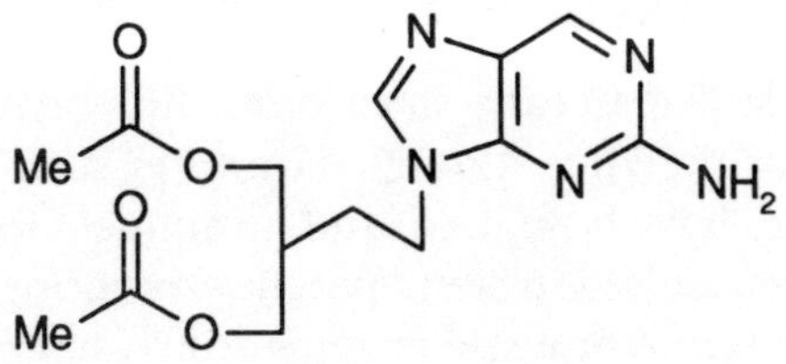

Ent-3TC/BW523W

FIAU

Famiclovir

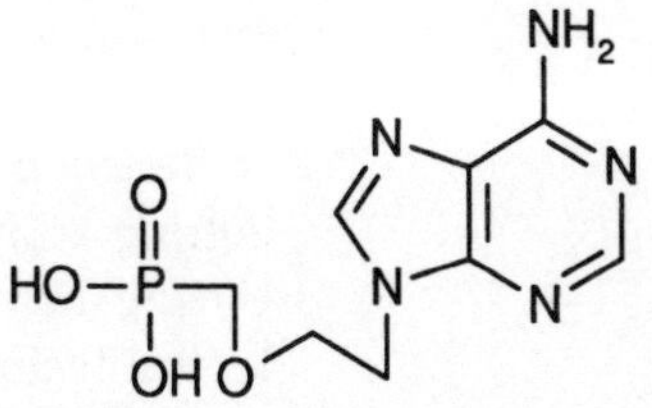

PMEA

Interferons

Chronic HBV infection can be treated with α-or β-interferons, although the basis of this is not true chemotherapy, but occurs by stimulation of protein expression. The HBV infection depresses the expression of major histocompatibilty complex (MHC) proteins on infected hepatocytes. These are targets for the virus specific CD8+ cytotoxic T-lymphocytes (CTLs). The expression of MHC class 1 proteins is stimulated by the interferon, which allows the CTLs to lyse the infected cells. The treatment can clear the infection completely under optimal conditions, though it is generally effective only in about 50% of cases. An interesting development in this area is the drug propagermanium, a polymeric organogermanium compound which is marketed in Japan for treatment of chronic HBV patients. This is believed to act by enhancing the endogenous production of interferons.

Immunization

A reasonable degree of protection has been provided using a passive immunization technique with human immunoglobulin which has a preponderance of anti-HB antibodies (HBIG). This has been superseded by an active technique whereby the natural defences are challenged by the injection of the surface antigens of HBV (HBsAg), which is produced by genetically engineered yeast cultures, thus stimulating the production of specific antibodies.

Other hepatitis agents

Hepatitis C virus

The causative agent of many non-A non-B hepatitis infections was identified in the 1980s and given the name hepatitis C virus (HCV). This enveloped virus, some 60-70 nm in diameter with an RNA genome of about 4,000 kDa, is morphologically related to flaviviruses, such as yellow fever virus. HCV, despite being a completely different virus, is transmitted in a similar fashion to HBV. Clinical implications of HCV infection, however, appear to be more serious; cirrhosis and hepatocellular carcinomas are more prevalent than with HBV infection. With increased knowledge about the virus, its transmission and epidemiology, HCV infection presents a major new target for chemotherapeutic research. An estimated 500 million people are infected with HCV, with up to 250,000 new cases each year; the geographical spread shows that between 0.5 and 8% of a native population might be infected, the highest incidence being in the Far East.

Research on HCV is still at an early stage; difficulties include the lack of a cell culture assay system for the virus (though surrogates such as yellow fever and bovine diarrhoea viruses have been used) and no animal models. Molecular biological techniques have resulted in some progress, resulting in the cloning and expression of the HCV serine protease enzyme which has been developed into an assay for testing potential inhibitors. Other identified gene products of future therapeutic potential include a metalloprotease, an RNA helicase and an RNA polymerase. Clinical conditions are being established whereby HCV infections are treated with interferons.

Hepatitis delta agent

Hepatitis delta agent, sometimes termed HDV or HδV, has a curious existence. It has a small virion (about 36 nm in diameter) with a single-stranded RNA genome of only some 600 kDa, which is too small to encode a full range of regulatory proteins. This incomplete virus, related to parvoviruses, is fully dependent on exploiting the biochemical processes present in an HBV infected cell; indeed, the viral outer coat is derived from an HBV antigen. HDV can therefore only co-infect with HBV, manifesting itself by exacerbating the effects of both the acute and chronic HBV symptoms.

Hepatitis E virus and beyond?

Hepatitis E virus (HEV), which might be one of a family, is a small (about 27 nm in diameter) virion, resembling calciviruses. Infection is by the enteric route and is spread in water, notably in large epidemics in India; its effects being particularly severe in pregnant women.

Hepatitis viruses A to E are certainly not the limit of the series, although the agent proposed as HFV appears to be a mutant form of HBV. HGV has been reported to be an RNA virus in the *Flaviviridae* family, but distinct from HCV and other known viruses.

9 The chemotherapy of respiratory viruses

Introduction

The general epithet 'colds and 'flu' is widely used to describe a variety of viral infections which inflame the respiratory tract. These very common diseases of man are widely self-treated in the West by anti-inflammatory and anti-pyretic drugs such as aspirin and paracetamol which may relieve the symptoms but have no effect on the virological parameters of disease. There are many different viruses which cause respiratory disease (Table 9.1) and it is inevitable that no single treatment is likely to be successful for all of them.

Table 9.1 Respiratory viruses of man

Virus	Family	Symptoms
Rhinovirus	*Picornaviridae*	Common cold
Adenovirus	*Adenoviridae*	Sore throats
Corona	*Coronaviridae*	Bronchial infections
RSV	*Paramyxoviridae*	Bronchitis Respiratory distress syndrome
Measles	*Paramyxoviridae*	Respiratory inflammation Rash
Mumps	*Paramyxoviridae*	Inflammation of parotid glands
Influenza	*Orthomyxoviridae*	Cough, malaise

The symptoms of the common cold—dry throat and sneezing caused by inflammation of the nose lining—are very familiar. They may be caused by a variety of different viruses (see Table 9.1 above) but are most commonly due to one of the rhinoviruses. The initial symptoms may be followed by secondary problems such as sinusitis or infections of the lower respiratory tract but in general they resolve in a day or two and the infection is cleared.

Influenza, which is caused by a discrete virus, is generally a more severe disease, being characterized by general malaise, sore throat, headache, muscular pains and a high temperature during the first phase and often a lingering cough afterwards. Morbidity is usually higher for influenza than for colds but once again the symptoms may resolve in a very short time and rapid recovery ensue.

Short duration self-limiting infections of this kind present their own problems when therapeutic intervention is required because the peak level of viral replication during which therapy might be expected to work often precedes the onset of the worst symptoms. The timing of therapy is crucial to successful treatment and must coincide with the period during which viral replication is at its maximum. Prescribing drugs quickly enough is therefore a serious problem. Alternatively, by continuous drug administration during periods of risk, such as during epidemics, the infection may be intercepted and prevented.

It is perhaps surprising in the light of these difficulties that a considerable amount of research has been done on the chemotherapy of colds. Several compounds have reached clinical trials and many more have been investigated in model studies. Reasons include the fact that the most common respiratory viruses, rhinoviruses, are comparatively easy to study in the laboratory and that there would be a high 'prestige' factor accruing to whoever was first in the field with a 'cure for the common cold'.

Rhinoviruses

Rhinoviruses, responsible for about two thirds of all common colds, belong to the picornavirus group, which also includes coxsackie and echo-viruses and those causing polio and foot and mouth disease. They are comparatively easy to grow in cell culture and for this reason have been extensively studied, as a result of which their biology and biochemistry is well understood.

Rhinoviruses are small RNA viruses with a single positive strand genome and an icosahedral protein coat built up entirely from four structural proteins arranged into 60 individual capsomeres. The viruses code for, and carry within their particles, an RNA-directed RNA polymerase and a protease, both of which are possible targets for chemotherapy. Also of potential therapeutic significance are the interactions between the structural proteins in the capsids which are responsible for maintaining the integrity of the virus particle. Like many other types of protein–protein interaction, these have the potential to be disrupted by 'foreign' molecules.

There is a large number of variations in the peptide sequences of the capsid proteins of rhinoviruses which results in a consequent variation in immunological properties. Over 90 different serotypes have been described, some with substrains within them. They are described by a number (e.g. RV-1B, RV-16, etc) which approximates to the order in which they were serologically characterized. One important consequence of this diversity is that the development of an effective vaccine is not practicable. Infection of an individual by one serotype, or artificial immunization, will result in future immunity to that serotype but not to other serotypes, so that re-infection is possible as soon as another serotype begins to circulate. With such a large number of possible serotypes, general immunity is not feasible.

Chemotherapy of rhinoviruses

Because rhinoviruses are easy to grow in cell culture monolayers they have been particularly suited to high throughput plaque assays and have been a target for mass screening of chemical compounds, both synthetic and natural product extracts. Many compounds active against rhinoviruses in cell culture have been identified by this means, some of which have been optimized by the classical procedures of medicinal chemistry into potent inhibitors of virus growth. The problem is that it is impossible to tell how the compounds are acting on the virus replication cycle or even whether what is being observed is a selective effect on the virus and is not some artefact arising from interference with one of the host cell functions. It is therefore customary to ascertain the mode of action of compounds discovered empirically in order to clarify the viral target and establish selectivity of action before they enter development.

It was found that inhibitors of rhinoviruses fall into two categories: those which are equi-active against all serotypes and those which are only active against some of them. The intrinsic value of a drug which only works against some of the target viruses is questionable but nevertheless several of them have proceeded as far as clinical trials.

Drugs which affect uncoating

Anti-rhinovirus compounds which are serotype selective must be binding to the exposed protein surfaces of the virus particles because it is only within these regions that the various serotypes of rhinovirus differ. Such compounds bind to the capsid proteins in such a way that the normal uncoating process which is associated with viral entry to the cell is interfered with, thereby preventing cellular infection. For one compound in particular, disoxaril, the interaction has been closely studied by X-ray crystallography for serotype RV-14 (Figure 9.1). The drug binds into a 'canyon' between two of the capsid proteins in such a way that the interaction between them is strengthened and disassociation does not occur under the normal conditions.

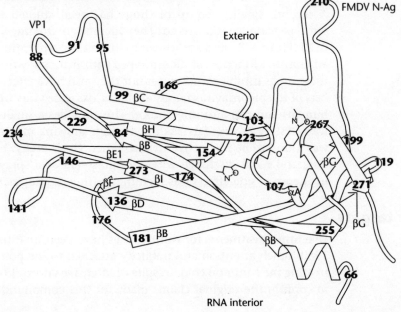

Figure 9.1(a)
A cartoon representation of viral structural protein VP1 from rhinovirus type 14, showing the position of binding of disoxaril.

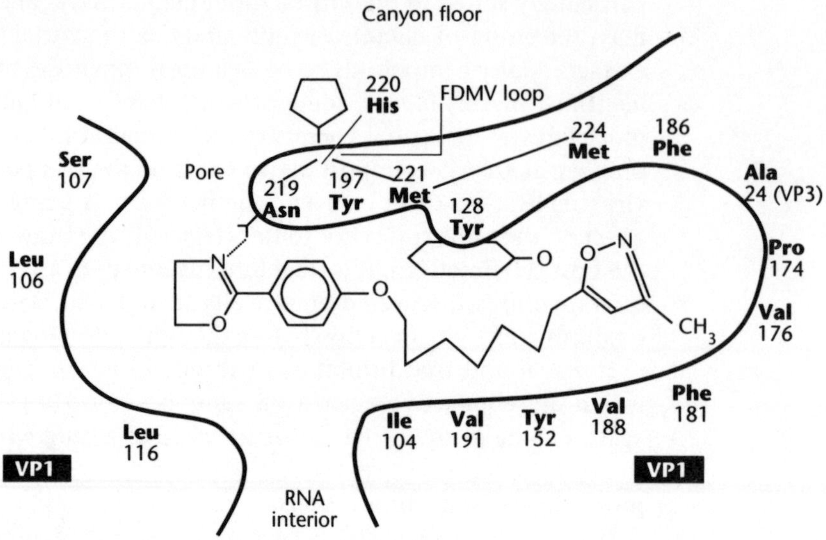

Figure 9.1(b)
Close-up of the 'canyon' region depicting detailed interactions.
(Reproduced with kind permission from *Science* (1986) **233**, 189.)

In spite of strenuous efforts, nobody has yet succeeded in discovering a single drug which stabilizes the capsid of all RV serotypes, though the spectrum of activity across the serotype range may vary considerably. Into this group fall BW683C, piradovir, Ro 09-0415, SDZ 89-365 and disoxaril itself (Figure 9.2). In clinical trials none of these compounds has shown any symptomatic benefit when given orally though, in some cases, if the compound is administered as a nasal spray it has been possible to demonstrate a reduction in virus shedding, especially when the drug is given at the same time as the infection.

Drugs which inhibit RNA polymerase

A second, smaller, group of drugs has been studied for which hardly any serotype variance is observed. They include enviroxime and the flavone, Ro 09-0179 (Figure 9.3), and are known to inhibit the replication of viral RNA which is an essential function of all serotypes with little known variation. Not only do these drugs inhibit rhinoviruses but they also have effects on some other members of the picornavirus family such as Coxsackie virus which are not specific to humans and can also infect mice. This provides a small animal model for efficacy testing *in vivo*. Enviroxime, though showing much promise at one stage, was found to be ineffective in man when given orally, but reduced viral shedding and nasal discharge when administered as a spray. In spite of this, the incidence of side-effects was such that further development was halted.

Other targets

Five other treatments for rhinoviruses have been investigated. During the mid 1980s much attention and publicity attached to the potential of vitamin C for treating the common cold. In spite of intensive efforts it has never been possible to confirm the original claims made for this compound as an antiviral and it

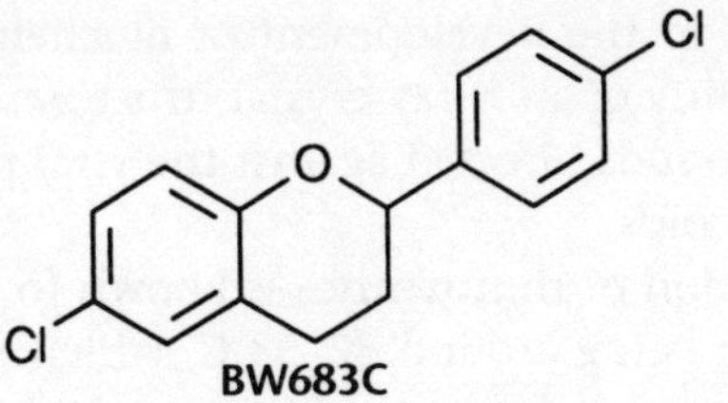

Figure 9.2 Drugs that show some sensitivity to RV serotypes.

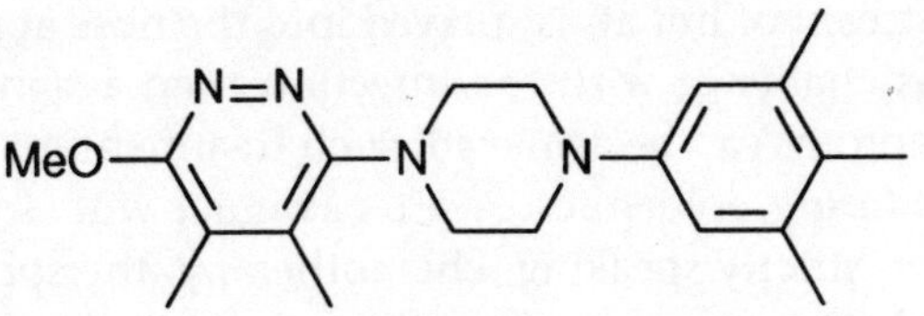

BW683C

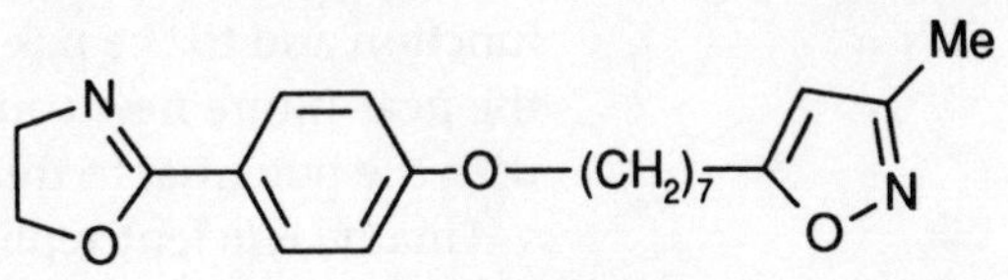

Disoxaril

Piradovir

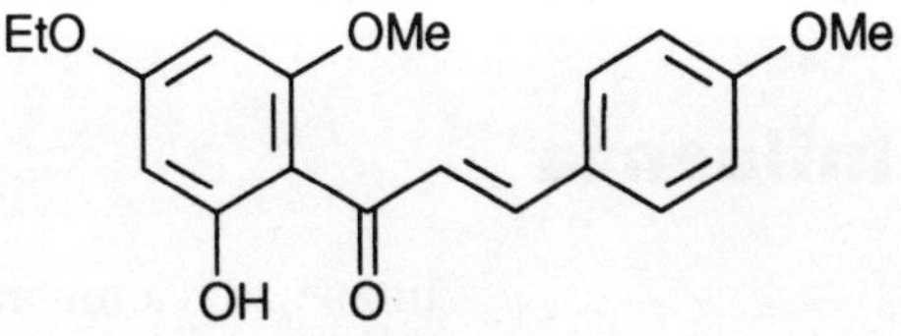

Ro 09-0415

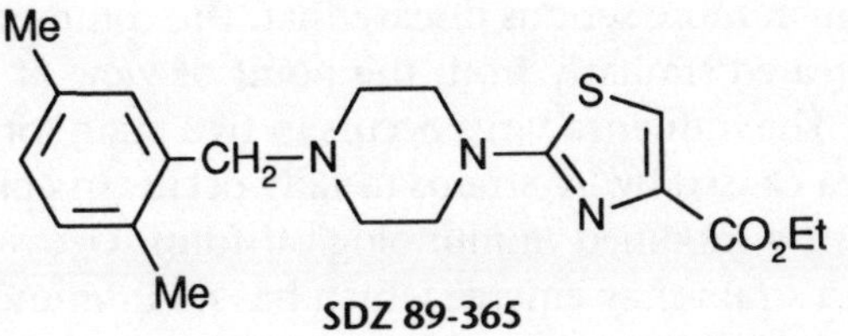

SDZ 89-365

needs to be dismissed from this consideration. Interferon (see Chapter 4) has been shown to be effective in man but the side-effects and economic factors are not really compatible with the treatment of a minor disorder. Zinc ions are known to inhibit rhinovirus growth in cell culture, possibly by interfering with nucleic acid metabolism by binding to metal chelating sites of the RNA polymerase. Zinc gluconate and fumarate salts have both been tried in humans but without success

The virally coded protease has recently received attention as a therapeutic target. The enzyme can now be produced in useful amounts by gene cloning and

Figure 9.3 Structures of enviroxime and Ro 09-0179.

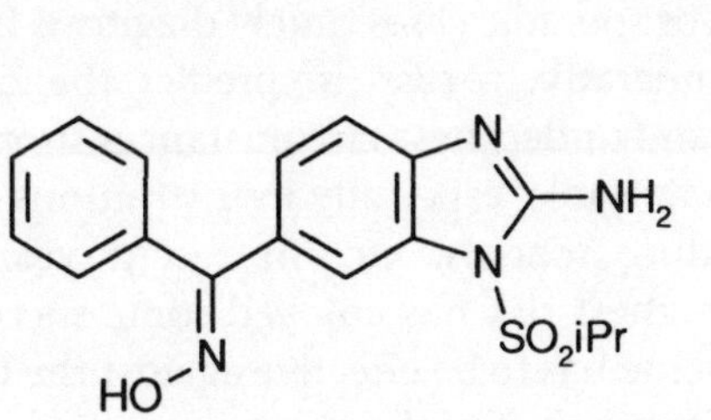

Enviroxime

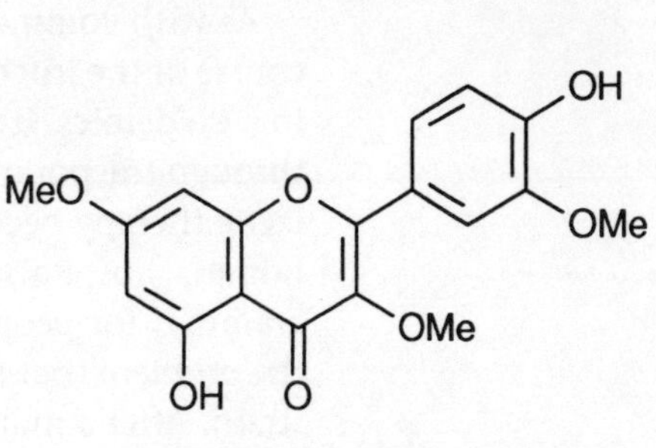

Ro 09-0179

expression, which has led to the development of biochemical assays for this function and to the possibility of an X-ray crystal structure. It is possible that in the near future new compounds directed against the viral protease will emerge with the potential to treat colds.

Finally, efficient replication of rhinoviruses is known to be dependent upon temperature, the optimum being around 32–33°C, which partly explains why colds are more prevalent in cold, damp conditions and why they infect tissues which are usually at a lower temperature than normal in the body. Studies have shown that if a stream of hot air is sprayed into the nose at frequent intervals at the same time as challenge with the infection then a significant reduction in incidence of symptoms can be achieved. Such treatment may be of value at periods of risk (i.e. during epidemics or after contact with infected individuals). Whilst this is not, strictly speaking, chemotherapy, the sprays have been commercially available in some countries and represent the nearest thing so far to a cold 'cure' that is available for use.

Influenza

Influenza is a much more serious disease than the common cold, although the two are often treated similarly from the point of view of prescribing symptomatic remedies. The influenza virus occurs in two main forms called 'A' and 'B' strains. Influenza caused by 'A' strains usually occurs in epidemics as a result of the emergence of a modified immunological form. Occasionally, due to gene recombination, a strain may emerge which has no immunological precedent at all and in such circumstances a world pandemic of great proportions may break out. When this happened towards the end of the first World War few people remained unaffected and worldwide mortality was more than 1% of the total population. Other serious pandemics occurred in 1947, 1957 and 1968–69. 'B' strains do not appear to undergo major immunological changes. Natural immunity therefore builds up in the population which results in generally milder symptoms, though they may be more serious for a primary infection in children. Apart from the lethal danger of influenza, which is greatest for the very young and the old or infirm, a tremendous amount of morbidity arises leading to loss of working time. Clearly a treatment for the disease would be highly desirable, giving economic benefits as well as relieving suffering.

As with common colds, to be of benefit therapy must be applied early in the course of the infection. For sporadic cases timely diagnosis is a problem but during epidemics it is comparatively easy to predict the course of infections through the population and under these circumstances short periods of prophylactic therapy become practicable, especially in institutions such as old people's homes, hospitals, boarding schools, etc. In recent years vaccination programmes for people at highest risk has enjoyed some success but suffers from the problem that the vaccine has to be effective against the currently circulating strain. After a major antigenic shift of the virus, protection will be poor or nonexistent.

Chemotherapy of influenza

Influenza is an enveloped RNA virus approximately 100 nm in diameter. It has two types of immunogenic spike embedded in its lipid envelope and a nucleo-protein core with a segmented genome. The principal features of the influenza virus are shown in Figure 9.4

There are eight pieces of RNA present in the genome, each of which codes for a single protein. These are: (a) two structural, or 'matrix', proteins (M1 and M2), concerned with viral assembly and disassembly; (b) two surface proteins (haemagglutinin, HA, and neuraminidase, NA) which form the spikes in the lipid envelope and are crucial to the process of entry to the cells, budding from cells and the maintenance of infectivity, and (c) four proteins concerned with the replication of viral RNA, an 'endonuclease' and a complex of three proteins which function as an RNA polymerase. All of these are viable targets for chemotherapy and have been studied as such. Due to the segmented genome the viral proteins are produced in their required form so there is no need for any post-translational processing by a protease.

Inhibition of attachment to the cells

The major surface protein of influenza is haemagglutinin, HA. Several hundred of these molecules are embedded in the lipid envelope of each virus particle. HA has been obtained in pure crystalline form and the complete three-dimensional structure has been determined by X-ray crystallography. The roles of the various regions of the protein, including the antigenic determinants have been

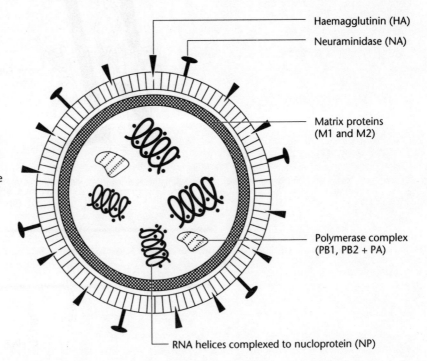

Figure 9.4 Principal features of the inflenza virus particle. The polymerase complex consists of three proteins (PB1, PB2, and PA). The nucleoprotein (NP) packs the RNA into a helical shape. There are eight pieces of RNA present in the core.

Haemagglutinin (HA)

Neuraminidase (NA)

Matrix proteins (M1 and M2)

Polymerase complex (PB1, PB2 + PA)

RNA helices complexed to nucloprotein (NP)

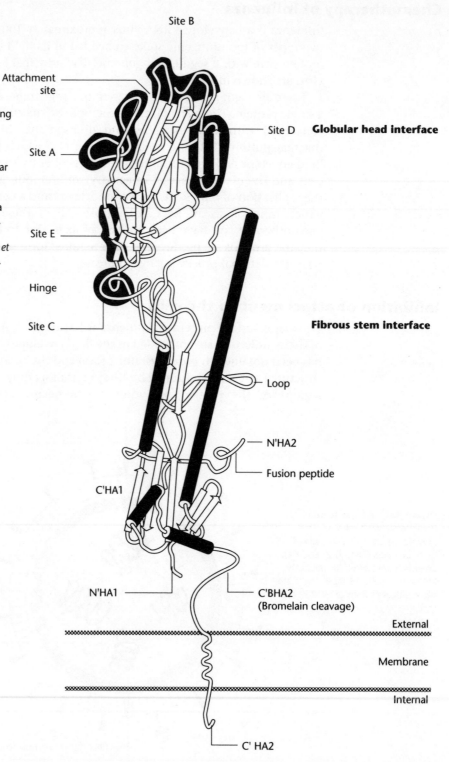

Figure 9.5 Cartoon depicting influenza haemagglutinin. Antigenic sites are shaded black. The neuraminic acid binding pocket is located near the head of the structure. In the membrane, three molecules are associated as a trimer for each surface spike. (Reproduced with kind permission from Wiley, D.C. *et al.* (1987) *Ann. Rev. Biochem.* **56** 365.)

Site B

Attachment site

Site A

Site D   **Globular head interface**

Site E

Hinge

Site C   **Fibrous stem interface**

Loop

N'HA2

Fusion peptide

C'HA1

N'HA1

C'BHA2 (Bromelain cleavage)

External

Membrane

Internal

C' HA2

identified with some accuracy. The haemagglutinin antigenic sites are depicted in Figure 9.5.

The function of HA is twofold. Firstly it brings about primary attachment to the cell through an interaction between a binding region which recognizes neuraminic acid residues and cell surface gangliosides which terminate in this residue (Figure 9.6). There are multiple interactions for any one virus particle and they have the effect of drawing the cell membrane round the virus so that it becomes contained in a vacuole within the cell. Secondly, as the pH falls in the cytosolic vacuole, the N-terminal region of HA undergoes a conformational change which initiates fusion of the viral lipid envelope with the cell membrane. Because of the specificity of the recognition requirement the neuraminic acid binding region of the protein is conserved in all strains of the virus and may therefore be considered as a viable target for therapy. Some simple analogues of neuraminic acid and some oligosaccharide sequences based on gangliosides have been found which do block the interaction but none has been developed further. The probable reason is that the interaction of neuraminic acid (Figure 9.7) itself with HA is fairly weak and analogues therefore become ineffectual because the large number of HA molecules present on the virus surface swamp any effect.

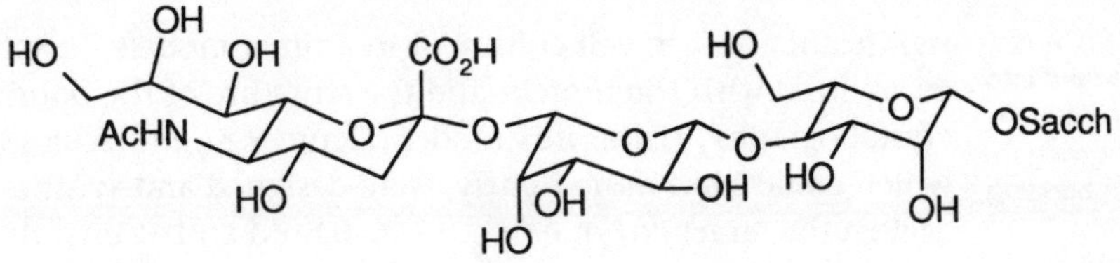

Figure 9.6 Terminal sequence of cell surface ganglioside.

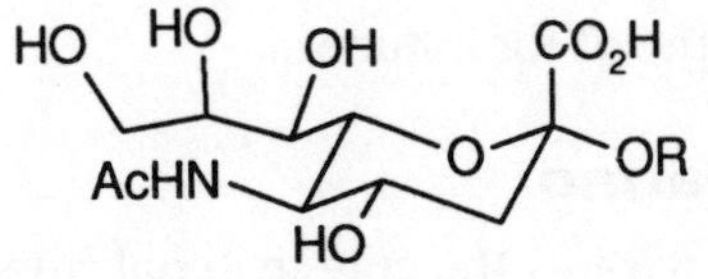

Figure 9.7 Neuraminic acid-based structures (R = Me, larger alkyl or non-cleavable glycosides), which block the binding of HA.

Inhibition of neuraminidase

Neuraminidase, NA, the other viral surface protein is present in much smaller amounts than HA. It catalyses the cleavage of terminal neuraminic acid residues from cell surface gangliosides, a process essential for the budding of new virus from infected cells and maintaining viral infectivity by releasing virus particles from non-productive interactions between HA and cell surfaces. Like HA, the protein has been isolated in a pure, crystallized form and had its structure determined. Also like HA, it has a region which binds the terminal neuraminic acid residue of cell surface gangliosides. In this case the terminal sugar residue undergoes a hydrolytic cleavage, following which the isolated molecule of neuraminic acid disassociates from the protein and frees the virus particle from its interaction with the cell.

Several inhibitors of NA had been discovered from testing programmes, including the neuraminic acid analogue (**1**), but they had not shown any

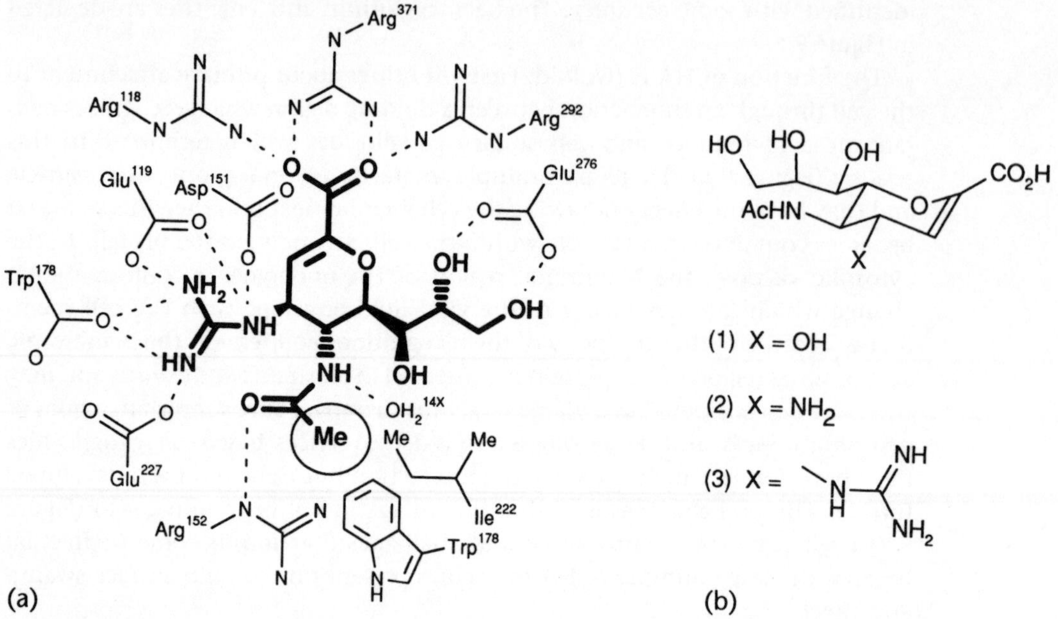

(a)   (b)

(1) X = OH

(2) X = NH$_2$

(3) X =

Figure 9.8 (a) Model of GGR167 bound into the neuraminidase active site, showing the key hydrogen bonding interactions. (b) Neuraminidase inhibitors.

significant effect in cell culture or in animal models. In spite of this (**1**) was co-crystallized with the protein and the structure of the bound complex solved by crystallography. Using this model (Figure 9.8), molecules (**2**) and (**3**, GGR167) which could bind more tightly, were designed and synthesized. (**3**) has activity against the target enzyme (IC$_{50}$ = 0.6–0.8nM and against the virus in cell culture of IC$_{50}$ = 5–14 nM). Furthermore, the compound was active against both 'A' and 'B' strains of virus, is not cytotoxic and is currently under development for clinical use as an aerosol treatment for influenza.

Matrix proteins—inhibition of uncoating

The sequence of events that releases the core material from influenza particles into the cytoplasm of the target cells is well defined. Attachment via HA molecules is followed by encapsulation of the particle into a vacuole (pinocytosis). Inside the vacuole a rise in pH triggers the fusion of the viral envelope with the cell membrane leading to primary uncoating and finally the matrix proteins are stripped off to release the viral nuclear material and functional proteins. The mechanism by which this process occurs appears to be by small positive ions passing through channels in the matrix protein, M2. The replication of influenza in cell culture can be inhibited by basic compounds such as simple amines and even ammonium salts, by neutralizing the pH change in the vacuole thus retarding the fusion event. Some compact aliphatic amines, showing potent activity against the virus, appear to block the ion channels in protein M2. Amantadine and Rimantadine, are the leading examples of this class of antiviral drug. They can be effective when given at the appropriate time, i.e. as early as possible, and have found limited clinical use against influenza 'A' infections. Their disadvantages are that they produce side effects due to penetration of the central nervous system, are only active against 'A' strains of influenza and resistance to the drugs develops very rapidly. Attempts to counteract these disadvantages have resulted in the more potent compound, ICI-130,685 (Figure 9.9), but it is still only active against 'A' strains.

Figure 9.9 Examples of compounds that appear to block the ion chanels in protein M2.

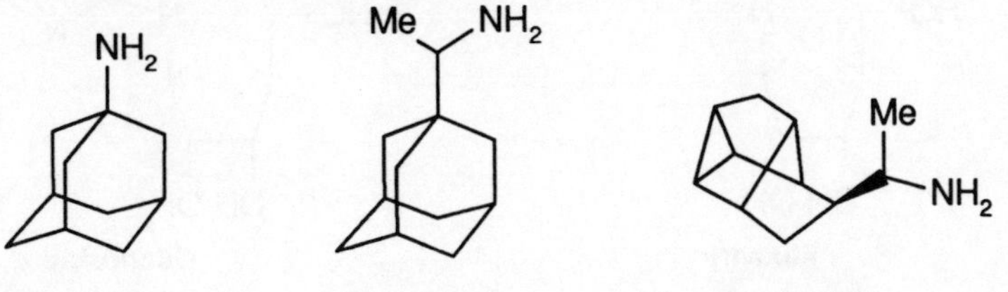

Amantadine   Rimantadine   ICI-130,685

Inhibition of the polymerase complex

Of many nucleoside analogues tested against influenza virus only a few examples have been found which inhibit the virus. They include carbodine, pyrazofurin and 3-deaza-adenosine (Figure 9.10), but have not found any clinical use because they are poorly selective for the virus, causing an unacceptable level of host toxicity. The toxicity problem is partly inherent in the fact that the active triphosphates of these compounds are produced by cellular enzymes and that there can therefore be no possibility of selective activation such as is seen for herpes viruses. Selectivity therefore has to depend upon the absolute preference of its triphosphate to inhibit the viral RNA polymerase complex rather than the cellular enzymes.

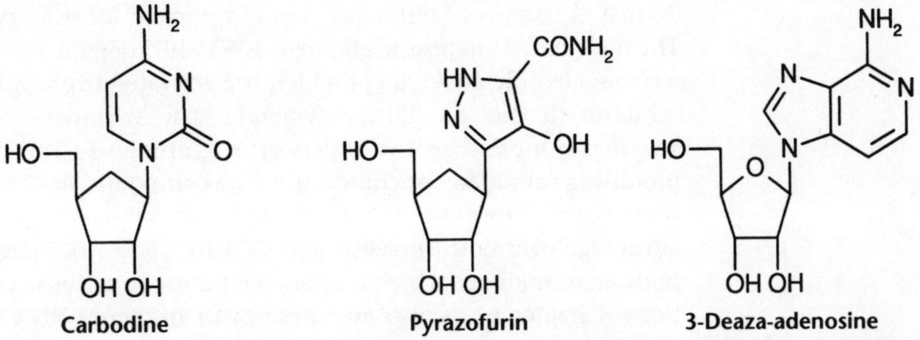

Figure 9.10 Anti-'flu nucleoside analogues.

Carbodine   Pyrazofurin   3-Deaza-adenosine

Given that the natural substrates of the RNA polymerase are ribonucleosides rather than deoxyribosides, one way in which this might be achieved is to incorporate sugars which mimic ribose into the nucleoside structure. One such compound is ribavirin, which can be regarded as a seco-guanosine, i.e with the 2- and 3-atoms of the purine ring system missing (Figure 9.11). It is interesting in that it has more than one mode of action. Cellular kinases convert it to the triphosphate form which is a partially selective inhibitor of influenza RNA polymerase. In addition however the monophosphate is an inhibitor of inosine monophosphate dehydrogenase—one of the enzymes responsible for maintaining cellular pools of guanosine triphosphate. The resulting depletion of the

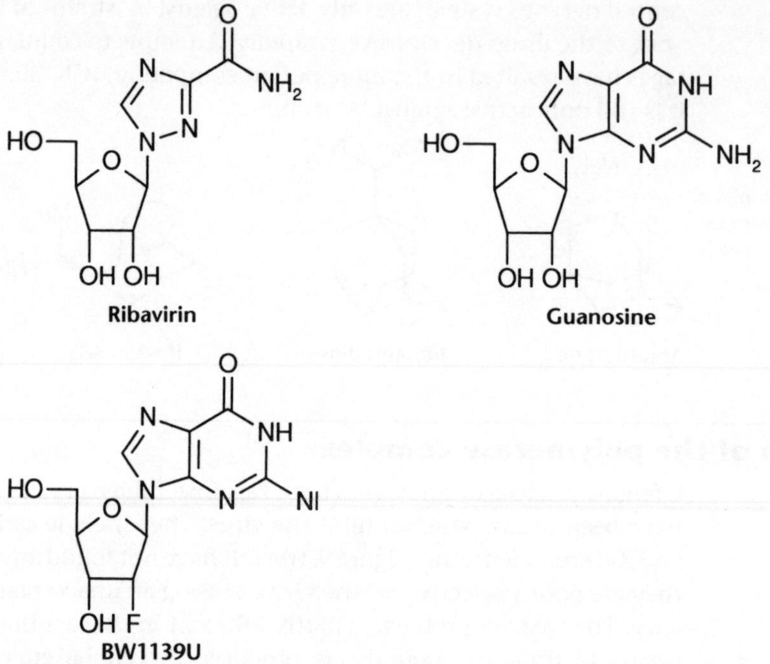

Figure 9.11
Guanosine and its
analogues ribavirin
and BW 1139U.

Ribavirin

Guanosine

BW1139U

pool of this essential building block for nucleic acids potentiates the activity of ribavirin. Ribavirin has disadvantages in that there are still some toxic side-effects and that it is ineffective unless given by aerosol directly into the respiratory tract. Nevertheless, as an aerosol, it has been shown to work in humans against 'A' strains of influenza, and is approved for use against RSV infections. The fluorinated guanosine analogue BW1139U (Figure 9.11), unlike ribavirin, is a ribonucleoside analogue of which the activated triphosphate form is only an inhibitor of the 'flu RNA polymerase. The selectivity compared with host enzymes is more than two orders of magnitude which is enough to make it a promising candidate for clinical use. The compound has been shown to be orally effective against both 'A' and 'B' strains in animal models, a further potential advantage over both ribavirin and GGR167. Foscarnet (Figure 5.25), the pyrophosphate analogue which is approved for use in serious cytomegalovirus infections (Chapter 6), is also an inhibitor of influenza RNA polymerase but not really selective in its action.

Miscellaneous

One further compound, still undergoing evaluation for potential use in man, should be mentioned in connection with influenza, which is the thiadiazole, LY 253963 (Figure 9.12). This very simple looking compound is orally active against types 'A' and 'B' by a mechanism which is not completely clear.

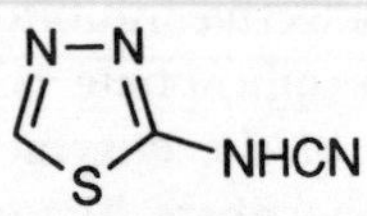

Figure 9.12 Structure
of LY 253963.

LY 253963

Chemotherapy of respiratory syncytial virus

As described in the preamble to this chapter, about 30% of so-called common colds are caused by viruses other than rhinoviruses. Chemotherapies for these infections have not yet been extensively studied except for respiratory syncytial virus (RSV). RSV can be a serious problem in infants, being a possible contributor to respiratory distress syndrome, a frequently fatal condition in neonates. It has been found that the drug, ribavirin is effective against the virus and its aerosol form has been brought into limited clinical use for such cases.

10 Future directions of antiviral chemotherapy

In considering the future potential developments in the treatment of viral conditions, it is first worth reflecting on what has been achieved so far. Antiviral chemotherapy is a young science, but nonetheless rapid progress has been made on a multidisciplinary front, which has led to the current range of therapeutics. This is especially true of AIDS and HIV research, where understanding of the fundamental biology and biochemistry of the virus has resulted in the discovery of more potential agents for treatment than for any other virus. This understanding has incidentally provided a great deal of insight into the biochemistry and functioning of viruses in general: exploitation of viral gene products, in addition to polymerases, as chemotherapeutic targets will surely bring progress against other diseases.

The most exploited area to date is that of nucleoside chemotherapeutics, a field which is by no means exhausted. There are sure to be new classes of nucleoside molecules with novel variations in the base or sugar portion which will produce active molecules. Research is continuing too on the phosphorylation behaviour of nucleosides and on the direct inhibitory potential of nucleoside phosphates and isosteres on viral polymerases. Systems for the intracellular delivery of nucleoside phosphates or better still, the design of more lipophilic isosteres, which have similar protein binding properties represent areas of rich potential.

The revolution in molecular biology will continue to provide new opportunities for intervention through the characterization of specific genes and gene products and then finding agents which disrupt their role. For a virus such as CMV, which posseses many gene products of unknown function, previously unappreciated targets for therapeutic intervention may be identified. This contrasts with hepatitis C virus, which cannot be handled yet in cell culture, where investigations into inhibitors of expressed viral proteins represent the only direct way forward other than the use of surrogate viruses. Proteins produced by expression systems can be used in screens for inhibitors of their specific function or can be crystallized and their structure solved, thereby providing a working model for inhibitor design. In some cases, culture cells can be genetically modified so that they can be infected with a particular virus, thus providing new and convenient assays for those which could not previously be assessed by conventional means.

In focusing on specific inhibition of viral replication by 'small molecule' chemical agents, the role of the immune system in viral diseases has not received detailed attention in this text on chemotherapy. Harnessing specific actions of the immune system, which itself has roles in the pathogenesis and/or persistence of many viral conditions, is likely to provide considerable therapeutic benefits. Agents which induce appropriate immune responses to viral invasions may offer an alternative or useful adjunct to specific chemotherapy. Vaccines, passive immunotherapy with γ-globulins and natural cytokines such as interferons, represent the contemporary immunological-based approaches: these are likely to be followed by other biological macromolecules and synthetic immunomodulators.

The majority of contemporary antiviral drugs are based on the inhibition of replication-associated biochemical functions, which prevent the production of progeny virions. This does not allow for any intervention during the latent phase of a viral infection, which is an important feature of many ubiquitous viruses such as those of the herpes family. Intervention during this poorly-understood phase has wide implications and potential; should a therapy be available that locks the virus in this dormant state, no viral gene products would be expressed, thus no effects would be observed. Inherently highly specific anti-viral therapies, acting in such a way, are amongst the many potential applications of oligonucleotides. As more is understood about chronic viral conditions, further progress may also be made in treating virus-associated cancers.

Even though many viral conditions can now be prevented by vaccination programmes or can treated with chemotherapeutics, there are many viruses pathogenic to humans which have not been investigated as medicinal targets. Some, such as Ebola virus have spectacular effects and dreadful symptoms, but their rarity reduces the commercial impetus for investigation. However, for these viruses and others, which will undoubtedly cause major outbreaks of diseases in the future, investigation of medicines already available is a likely first step, followed by research into specific targets by what will perhaps soon become well-established procedures. The means for directing research towards treatments for these diseases already exists and could be unleashed at any time given the economic or social impetus. In 1980, HIV was an obscure and ill-described virus which was thought to infect only African green monkeys. Things can change very rapidly indeed!

In spite of cutting-edge science, the reader should also have learnt why the Holy Grail of antiviral chemotherapy, a broad-spectrum therapy for the many agents responsible for the ubiquitous 'Common Cold', is probably unattainable.

Appendix 1

Viruses causing human diseases

Virus	Size (nm)	Family	Genome	Route of infection	Disease	Incubation period	Current Licensed/ Experimental Therapies
Vaccinia	250	*Poxviridae*	DNA	Skin Abrasions	Smallpox		Eliminated
Rotavirus	70	*Reoviridae*	RNA(−)	Ingestion	Diarrhoea		Rehydration
Polio	25	*Picornaviridae*	RNA(+)	Ingestion	Poliomyelitis	7–14 days	Vaccination
Rhinovirus	25	*Picornaviridae*	RNA(+)	Respiratory	Colds	1–3 days	Self–limiting
Hepatitis A	25	*Picornaviridae*	RNA(+)	Ingestion	Liver disease	3–5 weeks	Self–limiting/ Vaccination
Rubella	80	*Togaviridae*	RNA(+)	Respiratory	German measles	14–16 days	Vaccination
Yellow Fever	30	*Flaviviridae*	RNA(+)	Insect bites	Yellow fever	3–6 days	Vaccination
Hepatitis C	30	*Flaviviridae*	RNA(+)	Body fluids	Liver disease		Interferons
Bunya	100	*Bunyaviridae*	RNA(−)		Tropical fevers		None
Herpes simplex	150	*Herpesviridae*	DNA	Skin abrasions Sexual contact	Cold sores Genital sores		Nucleosides
Varicella zoster	150	*Herpesviridae*	DNA	Respiratory Reactivation	Chicken pox Shingles	13–17 days (years)	Nucleosides Nucleosides
HCMV	150	*Herpesviridae*	DNA	Respiratory	Various		Nucleosides
EBV	150	*Herpesviridae*	DNA	Respiratory	Mononucleosis	4–6 weeks	Nucleosides
Adeno	80	*Adenoviridae*	DNA	Direct contact Respiratory	Eye infections Cold symptoms		None used None used
Papilloma	50	*Papovaviridae*	DNA	Skin abrasions Sexual contact	Warts Genital warts		None None
Hepatitis B	40	*Hepadnaviridae*	DNA	Body fluids	Liver disease	10–12 weeks	Vaccination Nucleosides, Interferons
Corona	200	*Coronaviridae*	RNA(+)	Respiratory	Cold symptoms Bronchitis		None
RSV	150	*Paramyxoviridae*	RNA(−)	Respiratory	Bronchitis		Nucleosides
Measles	150	*Paramyxoviridae*	RNA(−)	Respiratory	Measles	13–14 days	Vaccination
Mumps	150	*Paramyxoviridae*	RNA(−)	Respiratory	Mumps	14–18 days	Vaccination
Influenza	100	*Orthomyxoviridae*	RNA(−)	Respiratory	Flu	1–3 days	Vaccination, Amantadine, Neuraminidase inhibitors
Ebola	800 x 40	*Filoviridae*	RNA(−)	Body contact	Haemorrhagic fever		None
HIV	100	*Retroviridae*	RNA(+)	Body fluids	AIDS	long	Nucleosides, reverse transcriptase/ protease inhibitors

Appendix 2

Antiviral chemotherapeutics: Major marketed products and selected late stage development molecules

Nucleosides

Structure

Generic Name   Acyclovir

Chemical Name   9-(2-Hydroxyethoxymethyl) guanine

Acronyms/Code Numbers   ACV; BW248U

Mode of Action   DNA polymerase inhibitor/chain terminator, activated only in infected cells

Target Virus(es)   HSV-1/2, VZV

Structure

Generic Name   Valaciclovir

Chemical Name   Acylovir-L-valine ester

Acronyms/Code Numbers   256W; val-ACV; BW256U

Mode of Action   Valine ester prodrug of acyclovir

Target Virus(es)   HSV-1/2, VZV

Structure

Generic Name   Famciclovir

Chemical Name   9-(4-acetoxy-3-acetoxymethylbut-1-yl)-2-amino purine

Acronyms/Code Numbers   FCV

Mode of Action   Oral pro-drug of penciclovir (PCV), selectively activated DNA polymerase inhibitor

Target Virus(es)   HSV-1/2, VZV

Structure

Generic Name	Lobucavir
Chemical Name	N/A
Acronyms/Code Numbers	Cyclobut-G; BMS 180,194
Mode of Action	Selectively activated DNA polymerase inhibitor
Target Virus(es)	HSV-1/2, VZV.

Structure

Generic Name	Vidarabine
Chemical Name	1-β-D-Arabinofuranosyladenine
Acronyms/Code Numbers	ara-A
Mode of Action	Non-selective DNA polymerase inhibitor, topical use only
Target Virus(es)	HSV-1/2

Structure

Generic Name	Idoxuridine
Chemical Name	2'-Deoxy-5-iodouridine
Acronyms/Code Numbers	IdU
Mode of Action	Non-selective DNA polymerase inhibitor, topical use only
Target Virus(es)	HSV-1/2

Structure

Generic Name	Trifluridine
Chemical Name	2'-Deoxy-5-trifluoromethyluridine
Acronyms/Code Numbers	N/A
Mode of Action	Non-selective DNA polymerase inhibitor, topical use only
Target Virus(es)	HSV-1/2

Structure

Generic Name	Ganciclovir
Chemical Name	9-(1,3-Dihydroxy-2-propoxymethyl) guanine
Acronyms/Code Numbers	GCV, DHPG
Mode of Action	Selectively activated DNA polymerase inhibitor
Target Virus(es)	CMV

Structure

Generic Name	Cidofovir
Chemical Name	9-(3-Hydroxy-2-phosphonylmethoxypropyl)cytosine
Acronyms/Code Numbers	HPMPC; GS 0504
Mode of Action	Phosphonate DNA polymerase inhibitor
Target Virus(es)	CMV

Structure

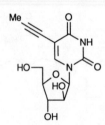

Generic Name	Sorivudine
Chemical Name	1-β-D-Arabinofuranosyl-5-*E*-bromovinyluracil.
Acronyms/Code Numbers	BVaraU; SQ 32756
Mode of Action	Selectively activated DNA polymerase inhibitor
Target Virus(es)	VZV

Structure

Generic Name	Zonavir
Chemical Name	1-β-D-Arabinofuranosyl-5-propynyluracil.
Acronyms/Code Numbers	882C
Mode of Action	Selectively activated DNA polymerase inhibitor
Target Virus(es)	VZV

Structure

Generic Name	Zidovudine
Chemical Name	3'-azido-3'-deoxythymidine
Acronyms/Code Numbers	AZT, BW509U
Mode of Action	Reverse transcriptase inhibitor/chain terminator
Target Virus(es)	HIV

Structure

Generic Name	Zalcitabine
Chemical Name	2',3'-dideoxycytidine
Acronyms/Code Numbers	ddC; Ro 24–2027
Mode of Action	Reverse transcriptase inhibitor/chain terminator
Target Virus(es)	HIV

Structure

Generic Name	Didanosine
Chemical Name	2',3'-dideoxyinosine
Acronyms/Code Numbers	ddI; BMY 40900
Mode of Action	Reverse transcriptase inhibitor/chain terminator
Target Virus(es)	HIV

Structure

Generic Name	Stavudine
Chemical Name	2',3'-didehydro-3'-deoxythymidine
Acronyms/Code Numbers	d4T; BMY 27857
Mode of Action	Reverse transcriptase inhibitor/chain terminator
Target Virus(es)	HIV

Structure

Generic Name	Lamivudine
Chemical Name	N/A
Acronyms/Code Numbers	3TC; (–)-BCH189
Mode of Action	Reverse transcription inhibitor/chain terminator
Target Virus(es)	HIV/HBV

Structure

Generic Name	Fialuridine
Chemical Name	1-β-D-2'-deoxy-2'-fluoro-arabinofuranosyl-5-iodouracil.
Acronyms/Code Numbers	FIAU
Mode of Action	Reverse transcription inhibitor
Target Virus(es)	HBV

Structure

Generic Name	Virazole
Chemical Name	1-(β-D-Ribofuranosyl)-1,2,4-triazole 3-carboxamide
Acronyms/Code Numbers	Ribavirin
Mode of Action	Inhibits, *inter alia*, viral RNA polymerase
Target Virus(es)	RSV, HCV and others *in vitro*.

Non Nucleosides

Structure

Generic Name	Saquinavir
Chemical Name	N/A
Acronyms/Code Numbers	Ro 31-8959
Mode of Action	Protease inhibitor
Target Virus(es)	HIV

Structure

Generic Name	Indinavir
Chemical Name	N/A
Acronyms/Code Numbers	L-735,524
Mode of Action	Protease inhibitor
Target Virus(es)	HIV

Structure

Generic Name	Ritinovir
Chemical Name	N/A
Acronyms/Code Numbers	ABT-538
Mode of Action	Protease inhibitor
Target Virus(es)	HIV

Structure

Generic Name	Nelfinavir
Chemical Name	N/A
Acronyms/Code Numbers	AG-1343
Mode of Action	Protease inhibitor
Target Virus(es)	HIV

Structure

Generic Name	Nevirapine
Chemical Name	N/A
Acronyms/Code Numbers	BI–RG 587
Mode of Action	HIV-1 reverse transcriptase inhibitor
Target Virus(es)	HIV-1

Structure

Generic Name	Delavirdine
Chemical Name	N/A
Acronyms/Code Numbers	U-90152S/BHAP
Mode of Action	HIV-1 reverse transcriptase inhibitor
Target Virus(es)	HIV-1

Structure

as

Na$^+$ O—P—O$^-$ Na$^+$

Generic Name Foscarnet
Chemical Name phosphonoformic acid , sodium salt.
Acronyms/Code Numbers PFA
Mode of Action Pyrophosphate mimic/DNA polymerase inhibitor
Target Virus(es) CMV

Structure

HO— OH
  H
HO
AcNH   O
          —CO$_2$H
  HN
      NH$_2$
  HN

Generic Name Zanamivir
Chemical Name N/A
Acronyms/Code Numbers GGR167
Mode of Action Neuraminidase inhibitor
Target Virus(es) Influenza

Structure

NH$_2$

Generic Name Amantadine
Chemical Name 1-aminoadamantane
Acronyms/Code Numbers N/A
Mode of Action Viral M2 ion channel blocker/maturation inhibitor
Target Virus(es) Influenza A

Structure

Me NH₂

Generic Name Rimantadine
Chemical Name α-methyl-1-aminomethyladamantane
Acronyms/Code Numbers N/A
Mode of Action Viral M2 ion channel blocker/maturation inhibitor
Target Virus(es) Influenza A

Structure

MeO — N=N — N—N — (3,4,5-trimethylphenyl)

Generic Name Piradovir
Chemical Name N/A
Acronyms/Code Numbers R-61837
Mode of Action Capsid binder
Target Virus(es) Rhinoviruses

Structure

N—O—(CH₂)₇— isoxazole Me

Generic Name Disoxaril
Chemical Name N/A
Acronyms/Code Numbers WIN-51711
Mode of Action Capsid binder
Target Virus(es) Rhinoviruses

Structure

benzimidazole NH₂, SO₂iPr, HO-N=

Generic Name Enviroxime
Chemical Name N/A
Acronyms/Code Numbers N/A
Mode of Action RNA polymerase inhibitor
Target Virus(es) Rhinoviruses

Glossary

Active transport The surface protein mediated process by which cells specifically take up essential compounds and many drug molecules.

Acute viral infection A productive infection where viral reproduction is occurring and infection is spread through lysis of infected cells or 'creeping' between neighbours.

Aetiology Study of the causes of diseases.

Anomers Isomers of carbohydrate molecules which only differ in the configuration at the hemiacetal carbon (carbon 1 or anomeric centre).

Antibiotic A substance used to treat infections caused by bacteria and fungi, through growth inhibition or destruction.

Antibody Complex proteins or immunoglobulins secreted by white blood cells which bind to antigenic foreign material and mediate its destruction by other components of the immune system.

Antigen A molecular species, usually of large molecular weight which can elicit an immune response in a host animal, through the production of antibodies.

Attenuated Describes a live virus used as a vaccine but which has been modified so that its pathogenic effects are minimized.

B cells Lymphocyte cells in the blood that are stimulated to produce antibodies by specific antigens.

Bacteriophage A virus that infects bacteria.

Bioavailability A pharmacokinetic term, which measures the fraction of an administered dose of a substance entering the general circulation. Oral bioavailablity can be affected by poor absorption from the GI tract and/or rapid metabolism in the first pass through the liver.

Capsid The protein capsule surrounding the nucleic acid core of a virus.

Capsomere A constituent protein unit of the capsid.

Cell culture The means by which groups of cells are maintained in a viable state away from their natural environment, usually in some form of laboratory glassware.

Chemotherapy The treatment of pathogenic organisms or control of cellular function by using of chemical agents.

Chronic viral infection The condition that occurs when a virus is found within the body, although it may not be producing symptoms, or even be in a dormant state.

Coat (of virus) The structure which surrounds the viral particle and protects it from the extracellular environment.

Core (of virus) The functional contents of a virus particle including its genome and essential enzymes.

cDNA Complementary DNA. DNA transcribed from an RNA template.

Chain terminator A molecule which acts as a substrate for a nucleotide polymerase enzyme, but lacks the free 3'-hydroxyl group which would allow the polymerization process to continue.

Chromosomes Components of mammalian cell nuclei which contain the genes.

Clone A group of individual organisms with an identical genetic make-up.

Creepers Descriptor of those viruses which pass between cells, rather than budding off or emerging after cellular lysis.

Cytopathic effect The changes observed in host cells as a result of infection by a virus, which includes induced cellular lysis or syncytia formation.

Dane particle The intact virion of hepatitis B virus.

DNA Deoxyribonucleic acid. The parent genetic material present in the chromosomes of all animals and in some viruses. A DNA molecule consists of a polymeric strand of individual deoxyribonucleotides, which is usually associated with a complementary strand with the characteristic double helix structure (dsDNA).

Double stranded dsDNA or dsRNA. Nucleic acids in which a sequence of nucleotide bases are hydrogen bonded to their complementary set.

Enveloped virus A virus which utilizes a lipid membrane derived from the host cell membrane as its outer coat, in which are embedded the viral surface proteins.

Epidemiology The study of how diseases are transmitted in host populations.

Expression A term used in genetic engineering when a gene produces a phenotypic effect, often in a foreign organism: e.g. in the production of a viral protein by a bacterium.

Family In viral families, a taxonomic classification suffixed *-viridae* above sub-families suffixed *-virinae*.

Gene A sequence of genomic nucleotides in DNA or RNA which carries the code for a single polypeptide which will ultimately constitute one protein or group of proteins.

Genetic engineering Experimental manipulation of DNA (or RNA) within organisms, to produce recombinant DNA which includes modified genes or genes inserted from a different species.

Genome The totality of the parent nucleic acid, encoding all the proteins of the particular species. Some viruses, unlike higher organisms, have RNA-based genomes.

Glycoprotein Glycosylated polypeptide protein molecule.

Glycoside A sugar derivative in which the hydroxyl group in the hemiacetal linkage is substituted, in one of two anomeric forms. Nucleosides are glycosylated, or specifically ribosylated, heterocyclic bases.

Glycosidase An enzyme which splits a glycosyl (sugar) unit from a structure, often specific to certain sugars (e.g. glucosidases split glucose units).

Haemagglutinin (HA) Influenza surface glycoprotein involved in host cell recognition and membrane fusion processes.

Immunoglobulin (Ig) One of a group of structurally related proteins, sub-divided into several classes that can act as antibodies.

Immunology The study of the way in which the body combats the invasion of foreign material by raising antibodies capable of facilitating the destruction or removal of that invading material.

Immunization The technique of artificially raising antibodies in a host to a potential pathogen by stimulating the immune system (vaccination) to respond to components of that pathogen. Passive immunization involves introduction of quantities of immunoglobulins which naturally contain levels of potential antibodies to certain invasions.

Immunocompromised Having impaired immune function as a result of infection by a virus such as HIV or as a result of immunosuppressive therapy in connection with organ transplants and certain autoimmune diseases, which suppress the immune system's actions.

Incubation period The period in-between primary infection and the onset of symptoms of an infectious disease.

Interferons A group of natural proteins produced by cells in response to certain stimuli, such as the presence of dsRNA, which have the ability to induce antiviral effects.

Isostere A molecule or part of a molecule of similar size and structure to another, usually with altered chemical and/or physical properties.

Kinase An enzyme which catalyses the phosphorylation of hydroxyl groups.

Latent state A state of dormancy achieved by viruses, such as those of the Herpes family, in which active replication and symptoms cease but the host remains infected.

Monocistronic A term that describes a genome in which all the proteins are translated from a single strand of nucleic acid as a single polyprotein.

Monotherapy Treatment of a patient's condition with a single drug: contrasted with combination or multiple therapy in which two or more drugs are used against the same target.

Morphology The description of the general shape and structure of virus particles.

mRNA Messenger RNA. The form of transcribed RNA used in translation.

Mutation A change in the amount or structure of the genetic material of an organism: it may be the change a single nucleotide (single point mutation) or multiple changes. Alterations in this genome will result in alterations in transcribed RNA and this in the make up of translated sequences in encoded polypeptides.

Negative strand A single-stranded nucleic acid in which the base sequence is complementary to the mRNA produced from it.

Neuraminidase (NA) Surface glycoprotein of influenza which has a catalytic function in trimming sialic acid residues involved in adhesion processes.

NNRTIs Non-nucleoside reverse transcriptase inhibitors.

Nucleic Acids DNA and RNA. Polymers of individual nucleotides, which have functions in protein synthesis and carrying a species' genetic information.

Nucleocapsid The core of a virus, containing the genomic nucleic acid plus structural and functional proteins.

Nucleosides The basic chemical building blocks from which nucleotides and nucleic acids are made. They consist of a heterocyclic base (adenine, guanine, cytosine and thymine/uracil) glycosylated on nitrogen by a pentofuranose sugar (D-ribose or D-deoxyribose).

Nucleoside phosphorylase An enzyme which catalyses the cleavage of the bond linking the base with the sugar moiety in a nucleoside or nucleotide.

Oligonucleotides Short nucleic acid sequences which are capable of binding to complementary sequences in natural DNA or RNA. These often have modified phosphate linkers to enhance stability.

Opportunistic infections Infections by some pathogens which exploit the weakening effect of a primary infection.

Pathogen Any foreign infectious agent which is harmful to the host animal.

Peptide A molecule containing two or more amino acids linked by peptide bonds (an amide linkage between amino and carboxyl functions).

Pharmacokinetics Study of the time course of drug concentrations in body fluids. The processes of absorption, distribution, metabolism and excretion (ADME) all contribute to a pharmacokinetic profile.

Phenotype The characteristic make-up of an organism related to its genome.

Pinocytosis The process by which cells envelop material and absorb it into the cytosol.

Plaque assay A convenient test in which the effect of a test drug on the cytopathic effect of a virus can be assessed in cultured cell monolayers.

Polycistronic A term that describes a genome in which the proteins are translated from separate pieces of nucleic acid as individual functional molecules.

Polymerase An enzyme which catalyses the transcription of nucleic acids from a complementary template.

Polymerase chain reaction A DNA amplification system, by which DNA sequences are enzymatically amplified, used in diagnosis and cloning. A similar process is possible for RNA.

Positive strand Nucleic acid which can function directly as mRNA in cell ribosomes, or has the same bases sequence as mRNA produced from it.

Pro-drug A derivative or related structure of an active drug molecule which has enhanced absorption and/or transport properties, which is metabolized to form the required drug, often at the site of action.

Prophylaxis Therapy which is intended to prevent acquisition of a disease.

Protease (or proteinase) An enzyme which is capable of cleaving peptide bonds in a polypeptide, usually with a degree of selectivity or specificity, thus producing two smaller peptide units.

Protein Macromolecules with a fundamental polypeptide structure.

Receptor A molecular feature of cell surface proteins which can be specifically recognized by other molecules (such as viral spikes).

Regulatory factors Proteins which control the sequence and amount of protein synthesis by interaction with nucleic acids.

Resistance The ability of a virus to become unaffected by a particular drug whilst at the same time preserving its pathogenicity.

Reverse transcriptase An enzyme which is capable of synthesising DNA from an RNA template. A characteristic component of retroviruses.

RNA Ribonucleic acid. The material which carries genetic information between the genome and the protein synthesizing ribosomes of host cells. Some viruses have a genome composed of RNA.

Serotypes Subtypes of a virus which are antigenically distinct from one another, usually arising from small changes in the surface proteins.

Spikes A descriptive term often applied to the surface glycoproteins of an enveloped virus which project from the membrane.

Strains Subtypes of a virus which contain a modified genome and consequently some minor phenotypic variation.

Surrogate virus A virus closely related biochemically to another, which is used in either or both screening or for *in vivo* evaluation of inhibitors of the second virus, where these would otherwise not be feasible.

Syncytia Multi-nucleated giant cells caused by some viruses, such as respiratory synctial virus, herpes viruses and HIV.

T cells Lymphocyte cells in the blood of several types that are responsible for cell-mediated immunity. Helper T lymphocytes are the specific targets for HIV infection, mediated through their CD-4 protein.

Transcription The production of RNA from a complementary sequence.

Translation The synthesis of polypeptides from mRNA templates.

Vaccine A foreign substance such an attenuated virus or viral surface antigen which elicits an immune response after inoculation, thus conferring immunity to infection from that virus.

Virion A complete single particle of a virus.

Further reading

General Texts

Collier, L. and Oxford, J. (1993) *Human Virology*. Oxford, Oxford University Press.

De Clercq, E. and Walker, R.T. (eds) (1988) *Antiviral drug development: a multidisciplinary approach*. New York, Plenum.

Dimmock, N.J. and Primrose, S.B. (1994) *Introduction to Modern Virology* Oxford, Blackwell Science Ltd.

Evans, A.S. (ed). (1989) *Viral Infections of humans: epidemilogy and control* New York, Plenum.

Fields, B.N., Knipe, D.M. and Howley, P.M. (eds) (1995) *Field's Virology*, 3rd edn. New York, Lippincott-Raven.

Fraenkel-Conrat, H. and Wagner, R.R. (eds), *The Viruses*, (multivolume series). Plenum, New York.

Galasso, G.J. Whitley, R.J. and Merigan, T.C. (1990) *Antiviral agents and viral diseases of man*, 3rd edn. New York, Raven.

Harper, D.R. (1994) *Molecular Virology*. Oxford, Bios Scientific Publishers.

Jeffries, D.J. and De Clercq, E. (eds) (1995) *Antiviral Chemotherapy*. Chichester, John Wiley & Sons Ltd.

Parker M.T. and Collier L.H., (eds) (1990), *Topley and Wilson's Principles of bacteriology, virology and immunity*. 8th edn, vol. 4, *Virology*, London, Edward Arnold.

Books, Reviews and Papers

The suggested reading list provided reflects a rapidly developing and expanding field. Whilst not intended to be exhaustive, it includes recent general reviews of many areas providing detailed bibliographies referring back to original papers. There are some chapters with few references: the material covered in these is covered in depth in the texts described above.

Introduction

Behbehani, A.M. (1991) The smallpox story: historical perspective. *American Society for Microbiology News*, **57**, 571–576.

Brown, F. (ed) (1990) Modern approaches to vaccines. *Seminars in Virology*, **1**, no. 1.

Fujinami, R.S. (ed) (1990) Mechanism of viral pathogenicity. *Seminars in Virology*, **1**, no. 4.

White, D.O. (1984) In: Melnick, J.L. (ed) *Antiviral Chemotherapy, Interferons and Vaccines*, Monographs in Virology. Basel, Karger.

Biochemical Assays for Viruses

Newton, C.R. and Graham, A. (1994) *PCR; Polymerase Chain Reaction*. Oxford, Bios.

Nucleoside Analogues as Antiviral agents

Chu, C.K. and Baker, D.C. (eds), (1993) *Nucleosides and Nucleotides as Antitumour and Antiviral Agents*. New York, Plenum.

Engel, R. (1977) Phosphonates as analogues of natural phosphates, *Chem.Rev.*, **77**, 349.

Herdewijn, P. and De Clercq, E. (1992) Future applications of oligonucleotides in antiviral and antitumoral chemotherapy. In: *Medicinal Chemistry for the 21st century*, IUPAC monograph, Oxford, Blackwell, 45–61.

Hertzberg, R.P. (1992) Agents interfering with DNA enzymes, Hansch, C. (ed) *Comprehensive Medicinal Chemistry*, vol 2. Oxford, Pergamon.753–792.

Hobbs, J.B. (1992) In: Purine and pyrimidine targets, Hansch, C. (ed) *Comprehensive Medicinal Chemistry*, vol 2. Oxford, Pergamon.299–332.

Holy, A. Travnicek, M. Snoeck, R. DeClercq, E. and co-workers. (1990) Acyclic nucleotide analogues. *Antiviral Research*, **13**, 295.

Huryn, D.M. and Okabe, M. (1992) Review: AIDS-Driven Nucleoside Chemistry. *Chem. Rev*, **92**, 1745–1768.

Propst, C.L. and Perun, T.J. (eds) (1992) *Nucleic Acid Targetted Drug Design*. New York, Marcel Decker. (Later chapters on oligonucleotides)

Symons, R.H. (1990) Self-cleavge of RNA in the replication of viroids and virusoids. *Seminars in Virology*, **1**, 117–126.

Townsend, L.B. (ed) (1988) *Nucleosides and Nucleotides*, (4 volumes). New York, Plenum.

Weintraub, H.M. (1990) Antisense RNA and DNA. *Scientific American*, **262**, 40–46.

Herpes Viruses

Cameron, J.M. (1993) New antiherpes drugs in development. *Rev. Med. Virol.* **3**, 225–236.

Collins, P. and Darby, G. (1991) Laboratory studies of herpes simplex virus strains resistant to acyclovir. *Rev. Med. Virol.*, **1**, 19–28.

Davidson, A.J. (1991) Varicella-zoster virus. *Journal of General Virology*, **72**, 475–486.

Elion, G.B., Furman, P.A., Fyfe, J.A., DeMiranda, P., Beauchamp, L. and Schaeffer, H.J. (1977) Selectivity of action of an antiherpetic agent 9-(2'-hydroxy-ethoxymethyl) guanine. *Proc. Natl. Acad. Sci. USA*, **74**, 5716–5720.

Furman, P.A., DeMiranda, P., St Clair, M.H. and Elion, G.B. (1981) Metabolism of acyclovir in virus infected and uninfected cells. *Antimicrob. Agents Chemother.*, **20**, 518–524.

Roizmann, B. Whiteley, R.J. and Lopez, C. (eds) (1993) *The Human Herpesviruses*, New York, Raven Press.

Rouse, B.T. (1992) Herpes simplex virus: pathogenesis, immunobiology and control. *Current Topics in Microbiology and Immunology*, **179**, 1–179.

Vere-Hodge, R.A., Sutton, D., Boyd, M.R., Harnden M.R. and Jarvest, R.L. (1989) Selection of an oral pro-drug of penciclovir. *Antimicrobial Agents Chemother.*, **33**, 1765.

Retroviruses (especially HIV and AIDS)

Clercq, E. De (1995) Perspective: Towards improved antiviral chemotherapy: Therapeutic strategies for intervention with HIV infections. *J. Med Chem.* **38**, 2493–2517.

Esnouf, R., Ren, J., Ross, C., Jones, Y., Stammers, D. and Stuart, D. (1995) Mechanism of inhibition of HIV-1 reverse transcriptase by non-nucleoside inhibitors. *Structural Biology*, **2**, 303–308

Huff, J.R. (1991) Perspective, HIV protease: A novel chemotherapeutic targets for AIDS. *J. Med. Chem.*, **4**, 2305–2314.

Katz, R.A. and Skalka, A.M. (1994) The retroviral enzymes. *Annu. Rev. Biochem.*, **63**, 133–173.

Larder, B.A., Darby, G. and Richman, D.D. (1989) HIV with reduced sensitivity to zidovudine (AZT) isolated during prolonged therapy. *Science*, **243**, 1731–1734.

Larder, B.A., Kemp, S.D. and Harrigan, P.R. (1995) Potential mechanism for sustained antiretroviral efficacy of AZT-3TC combination therapy. *Science*, **269**, 696–699.

Levy, J.A. (1994) HIV and the Pathogenesis of AIDS, Washington, ASM Press.

Martin, J.A., Redshaw, S. and Thomas, G.J. (1995) Inhibitors of HIV proteinase. In: Ellis, G.P. and Luscombe, D.K. (eds). *Progress in Medicinal Chemistry*, **32**, 240–281, Amsterdam, Elsevier.

Mitsuya, H., Weinhold, K.J., Furman, P.A., St Clair, M.H., Lehrmann, S.N., Gallo, R.C., Bolognesi, D., Barry, D.W. and Broder, S. (1985) 3'-Azido-3'-deoxythymidine (BW A509U): An antiviral agent that inhibits the infectivity and cytopathic effect of human T-lymphotropiv virus type III/lymphadenopathy-associated virus in vitro. *Proc. Natl. Acad. Sci. USA*, **82**, 7096–7100.

St Clair, M.H., Martin, J.L., Tudor-Williams, G., Bach, M.C., Vavro, C.L., King, D.M., Kellam, P., Kemp, S.D. and Larder, B.A. (1991) Resistance to ddI and sensitivity to AZT induced by a mutation in HIV-1 reverse transcriptase. *Science*, **253**, 1557–1559.

Hepatitis

Choo, Q-L., Kuo, G., Weiner, A.J., Overby, L.R., Bradley D.W. and Houghton, M. (1989) Isolation of a cDNA clone derived from a blood-borne non-A, non-B viral hepatitis genome, (HCV). *Science*, **244**, 359–362.

Clarke, B.E. (1995) Review: Approaches to the development of novel inhibitors of hepatitis C virus replication. *Journal of Viral Hepatitis*, **2**, 1–8.

Gerety, R.J. (ed) (1984) *Hepatitis A*. Orlando, Academic Press.

Hollinger, F.B., Lemon, S.M. and Margolis H.S. (eds). (1990) *Viral Hepatitis and Liver Disease*. Baltimore, Williams and Wilkins.

Hoofnagle, J.H. and Thape, D. (1986) Hepatitis and the hepatitis delta virus. In: Thomas, H.C. and Jones, E.A. (eds) *Recent Advances in Hepatology*. Edinburgh, Churchill Livingstone, 72–92.

Lemon, S.M. (1985) Type A viral hepatitis: New developments in an old disease. *N. Engl. J. Med.*, **313**, 1059–1067.

Reyes, G.R., Purdy, M.A., Kim, J.P., Luk, K-A., Young, L.M., Fry, K.E. and Bradley, D.W. (1989) Isolation of a cDNA from the virus responsible for enterically transmitted non-A, non-B hepatitis, (HEV). *Science*, **247**, 1335–1339.

Tiollaia, P., Pourcel, C. and Dejean, A. (1985) The hepatitis B virus. *Nature (London)* **317**, 489–495.

Respiratory Viruses

Al-Nakib, W. and Tyrrell, D.A.J. (1987) A new generation of antirhinovirus compounds. *Antiviral Research*, **8**, 179.

Cload, P.A. and Hutchinson, D.W. (1983) Inhibition of influenza RNA polymerase by pyrophosphates. *Nuc. Acids Res*, **11**, 5621.

Colman, P.M. and co-workers. (1993) Rational design of influenza sialidase inhibitors. *Nature*, **363**, 418.

Gilbert, B.E. and Knight, V. (1986) Biochemistry and clinical applications of ribavirin. *Antimicrobial Agents Chemother.*, **30**, 201.

Kubar, O.I., Brjantseva, E.A., Nikitina, L.E. and Zlydnikov, D.M. (1989) Importance of drug resistance in treatment of influenza with rimantadine. *Antiviral Research*, **11**, 313.

Smith, T.J., Vriend, G., Arnold, E., Rossman, M.G., Diana, G.D. and co-workers. (1986) Site of attachment of human rhinovirus 14 for antiviral agents. *Science*, **233**, 1286.

Sperber S.J., and Hayden, F.G. (1988) Chemotherapy of rhinovirus colds. *Antimicrobial Agents Chemother.*, **32**, 409.

Wiley, D.C., and Skehel, J.J. (1987) Structure and function of Influenza HA. *Ann. Rev. Biochem.*, **56**, 365.

Future Trends in Antiviral Chemotherapy

Boehme, R.E., Borthwick, A.D. and Wyatt, P.G. (1995) Immunotherapy of viral diseases, in antiviral agents section. Bristol, J.A. (ed.) *Annual Reports in Medicinal Chemistry*, **30**, 139–149, San Diego, Academic Press.

Braciale, T.J. (ed) (1993) Viruses and the immune system. *Seminars in Virology*, **4** no. 2.

Saunders, J. and Cameron, J.M. (1995) Recent developments in the design of antiviral agents. *Medicinal Research Reviews*, **15**, 497–531.

Updated Reviews and Papers

The following represent regularly updated volumes and monographs covering contemporary trends and developments

Bristol, J.A. (ed.) *Annual Reports in Medicinal Chemistry*. (1995) up to volume **30**, San Diego, Academic Press. Usually a chapter or more on antiviral chemotherapy.

De Clercq, E. (ed) (1993) *Advances in Antiviral Drug Design*, vol 1. Greenwich, CT, JAI Press Inc.

Ellis, G.P. and Luscombe, D.K. (eds) (1995) *Progress in Medicinal Chemistry* up to volume **32**, Amsterdam, Elsevier.

Galasso, G.J., Whitley, R.J. and Merigan, T.C. (eds) *Antiviral Agents and Viral Diseases of Man*, 3rd edn. New York, Raven Press.

Index